INCANDESCENT

Anna Levin

INCANDESCENT

We Need to Talk about Light

Saraband

Published by Saraband,
Digital World Centre,
1 Lowry Plaza,
The Quays, Salford, M50 3UB

www.saraband.net

ISBN: 9781912235315
eISBN: 9781912235322

1 3 5 7 9 8 6 4 2

Designed and typeset by EM&EN
Printed and bound in Great Britain by Clays Ltd, Elcograf S.p.A.

The author has endeavoured to contact every individual and/or a representative of
the organisations referred to in this book, and to reproduce the facts accurately and have
them checked by relevant readers. No liability can be accepted by the author or publisher
for any inaccuracies that remain, but please do contact the author with any queries or
corrections via her website: www.annalevinwriting.co.uk

For El

Contents

1

My Light Year

Where does any story begin? Does this one go back to the birth of light at the beginning of time? Or to humankind's own efforts to create light throughout the ages: from stone bowls of tallow and torches of bark and rush, through to whale oil, kerosene, gas lamps? Or to 1879, with Joseph Swan in his Gateshead conservatory and Thomas Edison in his New Jersey workshop, both refining incandescent light bulbs after years of trial and error?

Or does it begin in December 2008, in the shiny modern offices of the European Commission in Brussels, when representatives of the EU member states agreed to end the era of incandescent light?

I can only tell you my story. And if I scan time and space for the moment this strange tale of light collided with my life, I find myself in Birmingham in April 2013, feeling distinctly unwell.

I'm on a tree swing, leaning back and pulling at the rope, stretching my legs out so that as I surge up my feet reach over the treetops and into the clouds. The lawn and lake and huge trees appear and disappear, alternating with a blank expanse of white sky. But it doesn't make me feel any better. Why would it?

I slow the swing, touching one foot on the grass until it's still. Try breathing? I take deep, steady breaths into the side of my ribs, counting slowly. Or try that awareness thing? Look out from myself . . . I can see the lawn in front of me, Canada geese waddling across it; and I can feel the air on my face, the rough texture of the rope, the wooden seat beneath me. I can feel and see, but it's all at a distance.

It's no use. It still feels as if my head is floating somewhere above me. And I'm shaking. I look at my hand; it's not moving at all, but I feel like I'm trembling all over, and there's the strangest sensation, a kind of fizzy nausea.

I'm at Woodbrooke, a Quaker study centre in the heart of Birmingham. This stately Georgian manor house, with its boating lake and walled garden and arboretum, was once the home of the Cadbury family. It's my special place that I escape to when the hectic pace of family life leaves me ragged. It's like a deep well that I can come and drink from when I need to be replenished.

I live in Scotland and it surprises people that I travel to central Birmingham in search of inner peace. Why not go to Iona, with its abbey and turquoise water and white sand beaches? Or head for the Highlands and breathe the peace of the mountains? But I seem to find what I need in this oasis of big trees, communal dinners, warmth and friendship and honest conversations. These last years I've come for a writing course, which I find a revelation. It's about using writing as a tool, not to create a thing but to hear yourself think. To listen. I work as a freelance writer, specialising in wildlife and natural history journalism, and after more than a decade of that, I'd got used to shaping my writing to fit into boxes: target audience, style sheet, deadline and word count. Writing without those constraints is as exhilarating as freewheeling, and I always return home feeling I've gleaned some insight.

But this year something's wrong. Even the season seems wrong: it's April but chilly and grey and the trees are still leafless. It is as if winter's got this year in its teeth and it won't let go. It's so gloomy we've had the lights on in our normally sunlit meeting room. That's when it starts. As soon as the lights come on, a cold, creeping feeling comes over me, a tingling and burning on my scalp – my head feels like it's swelling and the shaking begins.

On this particular day, the lights are switched on at 3.30 in the afternoon. Our tutor is talking and suddenly I can't process the words I am hearing. There's a sensation like a balloon being blown up inside my skull and pushing against it. I have an overwhelming desire to get out of this room, to escape. Take no notice, I tell myself. Look out the window, concentrate – they're only light bulbs, after all. But some part of me is flinching and cowering from them, animal-like, trying to retreat into the furthest corner inside myself. The lights are those curly fluorescent ones, the new low-energy light bulbs. The ones that flicker on with a shudder then hang awhile with a dim, cold light before becoming slowly brighter.

I slip out and head upstairs to the sanctuary of my small,

Incandescent

cosy bedroom, where I curl up on the bed and examine the sensations in my body in turn. Once, I got an electric shock as a child when I touched a faulty lamp at my grandma's. I remember the thick, sickening jolt and the shaking after-wards – on and on, all down my arms and legs. Shaking inside, that's what this feels like. The nausea isn't like any-thing I've experienced before. It's not the urgent intensity of food poisoning, nor the queasy weight of a hangover; and it's nothing like the seeping, metallic constancy of morn-ing sickness. This is deeper, not located in my stomach but spread throughout my whole body . . . in my bones? My soul? An icky, ghastly *wrongness* that's hard to locate or identify. Somewhere, or something, between dread and poison.

I take the bulb out of the bedside lamp and study it. It looks innocuous enough, just an opaque white spiral. Clever, really – a strip light curled into a light bulb shape. A CFL: compact fluorescent light. It gives a poor, cold light, but they're supposed to be better for the environment.

Later, it's time for dinner. But I don't feel like eating or talking, so I go outside for a walk and a swing on the swing. I sit in the garden until the soft grey of the day turns to dusk and the weird feeling has faded a bit. I head to bed early and try to sleep, but an email I'd forgotten to reply to keeps buzzing around my head like a mosquito. Eventually I give in and pull a jumper over my pyjamas and wander over to a computer room in another wing of the building.

All along the way those curly bulbs, naked in their light fittings, seem to glare at me. I glare back, resenting them with a sudden rush of anger. They're ugly! They have no place in Woodbrooke's welcoming corridors, their austere light showing up every fleck of peeling paint, the way fluorescent strip lights in public toilets reveal every pockmark in your skin. Downstairs there's a new lounge overlooking the garden; it's big and spacious with comfy chairs, a wood-burning stove, plenty of books and magazines. But it's lit with a brutal white light more suitable for a dentist's practice. Does anyone else notice?

In the small computer room, I boot up the PC and I'm halfway through my email when the back of my head starts prickling and burning. On the wall behind me there's a big light in strips, reminiscent of an electric bar heater. I'm determined to get the email done, but my typing is clumsy, as if the link between my head and fingers is being interrupted. The back of my head is now actually stinging, like when you realise too late you've been in the sun too long.

I'm on autopilot. Press *Send*. Back down the grim-lit corridor into the darkness of my bedroom. Lie there drifting in and out of some heavy sleep-like state. Relentless thirst that water doesn't ease. The heat still burning, getting louder. Is heat loud? I sit up. 4am. The back of my head feels searing hot, even though it's several hours since I sat near that light. Sunburn's like this: it gets hotter afterwards. I stagger out of bed and soak a hand towel in cold water and wrap it around my head. Somehow I sleep like that. I dream I'm vomiting.

Morning brings the gentle relief of daylight seeping into my room. My tongue's too big for my mouth, which feels dry, but at least the burning has eased. I can even face breakfast. I walk around the boating lake first, the water's surface dark and full with the reflection of the trees all around. A sudden streak of astonishing turquoise shoots across it, shining like it's lit from within. Kingfisher. Its colours seem almost absurdly out of place among the drab hues of this dark April, as if it has been transported here from some brighter, warmer place.

I rush in to breakfast like some raving evangelist: *I saw it! The kingfisher! I saw a kingfisher!*

Then I try to explain to my group what's been going on, how every time I encounter those lights I start having these weird sensations. This gets mixed responses.

"Oh, I hate those lights! They make everything look like you're peering through the bottom of a pond."

"I don't think it could be the *light bulbs*."

"That's just a migraine. Haven't you had a migraine before? Don't worry, I get them all the time."

"I've changed all the lights in my house to those ones. We have to think of the environment."

And the cold white lights of the new lounge, I ask, *do they not bug you?*

"No. Why?"

"Oh yes, they're *horrible*! Makes me long to light a candle."

"What lights? I hadn't noticed."

So. The lights are as they are, but we respond to them completely differently, or barely notice them at all.

The sun makes a surprise appearance as we settle into

our morning's work and someone immediately shuts the blinds, saying they'd heard I was struggling with bright lights.

No, no – sunlight's fine! It's lovely, I try to explain. *No, it's not the brightness . . . it's the . . . it's the . . .*

What is it? I don't have the vocabulary to articulate my sense that what comes out of a CFL is a different kind of light, a different *stuff*. I can physically feel it.

I stick with the course until the afternoon, when some of the older participants need the lights on to read. Then I disappear to my room. By the evening I'm in the mood to talk and I seek out my course mates. My friend Brian and I venture out to stroll through the streets of Bournville, once a village built by the Cadbury family for the chocolate factory workers. We walk on and on through the pretty village-inside-a-city, with its tree-lined streets, chocolate-scented air and the orange glow of street lights.

Back in Woodbrooke's big dining room at ten in the evening, we settle for tea and biscuits. The left side of my head starts burning again, closest to where the CFL lights are high on the wall. Brian just gets up and puts the lights out for me and we sit drinking tea in the semi-darkness. Anywhere else this would seem strange, but one of the "Friends-in-Residence", who has a warden's role, comes in to check on us and just smiles: "Are you peaceful in the dark?" she asks. It's kind of ironic, we agree, having trouble with light here. Quakers talk a lot about light, but not in the sense of lamps and bulbs. It's been a constant spiritual metaphor since the movement's seventeenth-century beginnings, when the founder George Fox preached about "the inward light" available to every person that guides us and pushes us to action. We talk of a light to live by, a light that shows us the way, an inward illumination that reveals our darkness. We "hold each other in the light" in difficult times.

The next day I'm homeward bound at Birmingham's New Street Station. I'm browsing the sandwiches in M&S, deep in decision-making, when my head begins to burn and swell again. I look up. Big fluorescent strip lights like planks are hanging high above. I pull my coat over my head and dash to the subterranean platform for the Edinburgh train. The many miles north pass in a blurry daze, interspersed with pangs of regret that I never did get that sandwich.

When I get home I sit on the doorstep and breathe it all in. From my front door, central Scotland stretches before me in layers of time and space: east to the pale blue smudge of the Firth of Forth, and over to Fife; straight across to the Ochil Hills; and west to Stirling and the mountains beyond. And sometimes, on a clear day like this, the mountains peeking behind them, and more behind, on and on. The sun gleams on a faraway rectangle of pale yellow: the Great Hall of Stirling Castle, restored to the colour of its six-teenth-century origins. Downhill to the east is the surreal city of Grangemouth Oil Refinery, all chimneys and towers and smoke and flares. In the decade we've lived here, we've watched white wind turbines dot the hillsides, and two giant steel horse heads – the Kelpies sculpture – rear up over the urban sprawl of the central plain. It's a vast, erratic patchwork of eternity and modernity.

But inside the house something has shifted. I flinch when anyone switches the kitchen lights on – they feel too bright and harsh, and make me irritated and snappy with anyone around me. They're just those halogen spotlights that you get in many kitchens, and I've never really noticed them before. Which makes me wonder: am I just worrying, noticing too much, and is this somehow making me over-sensitive?

But no, when I head upstairs to my attic office my mind is already absorbed in the article I have to finish, and I'm not even aware of switching the light on at the top of the stairs. It's an LED with a blue-ish hue and as soon as it comes on a thick wave of nausea sweeps through me. After about half an hour at the computer screen, my face begins to prickle. It seems that my sensitivity to all sorts of light has been heightened. The incandescent bulbs that we mostly use are fine, though. And sunlight, of course – thank God.

"It's just a phase. Don't worry so much," say friends and family.

"It's just some kind of migraine – it'll pass."

"You were wrung out, over-tired. Get some rest and you'll be back to normal."

I hope they're right. And anyway, at long last it is May, my favourite month and the peak of my Scotland year: islands full of seabirds and skies full of hope. And true to form, as soon as May picks up the reins, the trees turn green and shimmer like they're bursting into song, the hedgerows froth with hawthorn blossom, the woods are filled with birdsong and the days stretch on and on, as if they're savouring it all too. I walk with my dog, drinking in the colours, feeling relief. We've barely had an electric light on for weeks, and I realise the fog in my head has cleared. The computer isn't affecting me anymore either. I think I'm okay!

•

My mobile phone battery is on the blink so I head into town on a Saturday to see if I can find a new one. The short queue is taking forever. As I wait, I'm struck by a sudden, ferocious thirst, and my lips start tingling. I glance up and see bare CFLs shining right above me. But it's a bright day and the small shop is full of sunlight, too. I decide to stick it out and get my phone fixed. Then comes that animal urge again, telling me to get out now. But could I be imagining or even creating that compulsion by remembering it from before? I wait, get my phone sorted and head out into the sunshine and weekend shoppers.

I'm halfway down the high street when I feel the lurch – an abrupt jolt and a distance from the world around me. It's a bit like having a couple of drinks on an empty stomach – but that's pleasurable and this is anything but. I get into the car and find myself wondering which side of the road is the left. I can't seem to figure it out. There's a blue car on the street ahead and it's moving towards me as if in a dream. So I decide to leave the car in town and get the train home. I need to buy a ticket and the guard is looking at me expectantly, but no words are available. It's like my vocabulary is packed away under something woolly and I can't retrieve it.

"Are you *sure* this isn't just anxiety?" my GP asks me kindly. "Could you have heard or read something about these lights and worry now when you see them?" I had dashed into the doctors' surgery with my coat over my head to shield me from the CFLs in the low ceiling of the waiting room. No wonder she's looking concerned. I shake my head. I'm well acquainted with the shape-shifting creature that is anxiety: sometimes a small, clawed rodent scratching at me insistently; sometimes a horse galloping on my chest, snorting and sweating, nostrils flaring; other times, at night, a snake that wraps around my limbs and torso and gradually tightens. I know the whole damn menagerie all too well, and this is something different: physical sensations that I've never experienced before. And besides, I tell her, there have been plenty of times that I'd no idea the lights were there and wasn't remotely worried until the symptoms began.

I have so many questions. *What are these lights? What are they doing to me? And how? And what I really don't get*, I say, *is why now?* These lights have been around for a few years now and most of my friends' houses have them. I've been to Woodbrooke five times in recent years and I have previously noticed the bulbs and their ugly pallor, but I've never had any reaction before. What's changed? Will it pass? Will I get better?

She looks at me sadly and shakes her head, deep in thought. No, it's not a migraine, she confirms, nor an allergy as such. Any previous issues with lights? *No.* Flicker? *No.* Any history of migraine? *No, not even a headache.* Problems with computer screens? *No, I work all day on a computer.* Any other possible connected health issues? *Nope.* I don't know, she says with a sigh, we don't know. These technologies are new, and medicine hasn't necessarily caught up yet. She will refer me to any specialist who may have an insight: dermatologist, neurologist, psychologist . . .

"Oh, and the street lights!" she says suddenly. "They're changing the street lights!"

I nod. I feel hollow and empty inside, but with a sense of creeping apprehension.

I take my questions to Google instead, that night and over the next weeks. There is a surprising amount of discussion out there. Incandescent lighting has been phased out and banned by the EU for environmental reasons, forcing a change to low-energy lighting, including CFLs and LEDs. And I'm not the only one having problems with the new lighting. A woman in Canada writes a blog about the effect of CFLs on her:

> Even a few minutes in these lights will have me unable to think, pressure in the front of my head, feeling weak . . . and if a few more minutes, headaches and nausea. Much longer and I do throw up . . .

I read of a man in Vancouver who collapses immediately when CFLs are switched on. *Something* must be disrupting the messages from his brain to his limbs. A lighting designer describes the light of CFLs as only fit for morgues and dentist waiting rooms. That makes me smile. Yikes, someone sitting under CFLs finds her scalp starts actually bleeding. I discover that CFLs will soon be replaced by LEDs – the newest form of lighting – but they are causing problems too: eye pain, headaches and migraine. A campaign group says that people suffering from lupus, an autoimmune disease, must have access to incandescent lighting – none of the newer bulbs are tolerable.

The position of environmental campaigners has been clear-cut in recent years. As a Green MEP, Caroline Lucas led the parliamentary group that promoted the ban on incandescent lighting. "Getting rid of incandescents is a no-brainer," said the WWF, as the ban will save "15 million tonnes of CO_2 by 2020" . . . equivalent to Romania's annual electricity usage. WWF called for the shops to be cleared of incandescent bulbs, and for EU anti-dumping duties to be relaxed so we could get more CFLs rapidly imported from China. Greenpeace launched a campaign to promote CFLs in India, naming "climate villains" who produced incandescent light bulbs, and got schoolchildren to make a human formation spelling "Ban the bulb" to promote the switch to CFLs. Greenpeace even crushed 10,000 incandescent bulbs in Berlin with a road roller – "Incandescent light bulbs are so dangerously inefficient that it is better to smash even brand new ones than to use them," the organisation announced.

The next Saturday I take my daughter to another two-year-old's birthday party, and it's fast descending into a black comedy. They don't *get* pass the parcel yet. The enticing parcel is handed to a young boy, who looks surprised then begins to tear it open. No, no, no – the mum rushes in, gesturing for him to pass it to the girl beside him. He begins to cry, and so it goes on.

I'm enjoying this scene (possibly more than I should be) when a sickly dizziness creeps over me. I look around and spot a lamp in the dining room, so I slip out and sit on the garden wall. But my daughter hangs at the kitchen door, arms outstretched, wanting me to come back and play musical statues.

When we get home, I find that worry is settling into my stomach like something leaden, dragging me downwards. I need space to think. I go out with the dog and walk, until a sudden wrongness floods my limbs and body. Damn it. I'd been watching the sky over the west and hadn't realised the street lights had come on behind me with a cold, fluorescent glare.

That night, I collapse into bed, only to wake at four in the morning, retching and with a raging thirst. I've been dreaming of being forced to drink paint. Creamy emulsion, magnolia, being poured down my throat. I sit up for a drink of water and see the daylight sneaking around the curtains. May is waiting outside, all bright greens and blue sky. I love this month so much that I usually don't want it to pass. But this year there's a darker note. Winter-fear is already beginning to grip me, casting a shadow over my shining month.

Other than school run hellos, I don't feel like seeing

anyone for a while. Life has shifted on its axis and I need time to adjust. Some kind of claggy curtain has come down between me and the world and I need to process what's happening on my side of it before I find a way to reach through. To those looking through, I appear no different; but on my side I'm curled into a ball, weighed down by what-ifs and relentless, niggling questions . . .

What was actually wrong with incandescent light bulbs? Do we really *have* to have CFLs or LEDs in order to save energy? Is it a moral imperative? Greenpeace, Friends of the Earth and the WWF say so, and they're all organisations that I have always wholeheartedly supported before. I don't get why incandescents were *banned*. We don't, as a rule, ban things in this country, even when we should. Normally, taxes, financial incentives or publicity campaigns are used to bring behavioural change in our shopping habits. Consumer choice is paramount in every other sphere of life. Often absurdly so. How come we can choose to have as many gas-guzzling cars as we like, fly around the world whenever we wish, leave the lights and heating on all night, but we can't choose a light bulb?

I turn to Google again . . . Apparently, fluorescent light has to contain mercury – that's how it makes light. People ask me whether *that* could be causing my problems: can it escape? Does it leak out? I don't think so. But I'm intrigued, and confused. I read that mercury is a potent neurotoxin, particularly dangerous to children and the developing foetus. Each CFL bulb contains a tiny amount, about 4mg of mercury. Is that a problem?

No, says the "mercury paradox": coal-fired power stations release mercury into the atmosphere, and therefore using mercury in bulbs actually results in less overall mercury, as we use less electricity with new bulbs.

Yes, says the charity Toxics Link: we're dealing with one of the deadliest toxins in everyday use and small quantities can have major impacts on our health. Toxics Link is based in India, where the mercury content of CFLs seems random. The average bulb was found to contain 21mg; the highest was 62mg.

And is there a risk if a CFL bulb breaks?

No, says the UK government. NHS guidance is calm: a breakage is "extremely unlikely to cause problems to your health", but it's recommended to take "sensible precautions" when clearing up afterwards.

Maybe, says the US Environment Protection Agency, which suggests more severe precautions: get people and pets out of the room immediately, shut off air conditioning, do not hoover.

Yes, says Toxics Link, as most bulbs are not disposed of properly and some countries do not have the necessary infrastructure to deal with their disposal, meaning mercury can enter the water cycle and food chain.

I'm getting obsessed. I'm Googling when I should be working, when I should be reading stories to my children. This story has so many strands – physics, physiology, politics, ideology, ecology – it's growing arms and legs, tentacles even. The monster that is forming from my searches is beginning to resemble a giant squid. When I'm trying to

get to grips with one vast tentacle, I find another creeping around me, thick and muscly, with suckers the size of
dinner plates.

My neighbour Anne is made of no-nonsense warmth
and straight-talking clarity. She knocks on the door one
day, snatches me from the squid's grip and hauls me away
for a day out. It's the Open Studios weekend, when artists
throughout the area open their homes to share their work.
Anne drives us towards Stirling and along a bumpy farm
track until we reach a converted barn. Inside the lights are
okay – halogen spotlights, I reckon – and it's bright with
sunlight anyway. We browse slowly and come to some glass
boxes containing curious sculptures of sphagnum moss
and curls of silver birch bark, lit from below. I'm intrigued.
The artist is on hand to explain.

"I'm obsessed with paper, and paper is light," she says
earnestly, and continues to tell me how paper is, essentially,
cellulose – plant matter – which is made by photosynthesis. So ultimately it's all about light . . . I chew this over
downstairs, along with home-made cakes and tea. Did I
pull this conversation to me through some magnetism of
obsession, or is there no escaping light?

Not that I'd want to escape June daylight. It lingers
over the mountains long into the evening, lighting up
Stirling castle and shining on the curves of the Kelpies.
I want to suck it all in, gather its abundance and store it
in me somehow to sustain me through the dark nights to
come. At some stage I try to shut it out to attempt to get
the children to sleep. Their room has curtains and blackout

blinds, but it's a losing battle – June light sneaks around the edges of the blinds, playing peekaboo with my giggling daughter.

•

School term ends in the last days of June, and we decide to head north as far as we can go to soak up the light. Our first overnight stop is in the Black Isle on the eastern shore and once the children are tucked in to their campervan beds, we settle down to enjoy a whisky. But once again the light whispers mischief – at 9pm it's as bright blue as a summer's day, albeit with a sweeter note, a hint of evening magic. The kids are giddy with daylight and freedom, peeking out of the window to the view so full of promise.

In the morning we head down to Cromarty Point, a long spit reaching into the North Sea where bottlenose dolphins often pass close by, fishing on the rising tide. By sheer luck we've timed it well and it's not long before a dark fin appears far out in the shining sea. Then another, closer by. Then dolphins are streaming in towards us from three directions, and my son keeps up a running commentary: "There! There! *There!*" I squeeze my daughter's hand tight. Her first dolphins.

I could watch forever but it's time to continue north again. At 11pm we're on the beach in Durness on the far north coast of Scotland, paddling in the blue, blue sea and jumping over little waves. I tell my daughter that it's actually night-time, way past bedtime. Just not dark! "Heh?" she says, one side of her nose wrinkled up in bafflement. Then she giggles. For surely night is dark, and day is light, and Mummy is talking nonsense.

We meander slowly back to the waiting campervan – it's not getting darker as such, just softer and dreamier. The colours are all still there – pale turquoise water, pale gold sand, rocks and stones in every shade of pink and mauve and grey, but muted somehow. I pick up a gentle pink stone and turn it over in my hand, wondering what its "real" colour is, and how it changes through the day and the year as the light changes.

One last precious night on this northern shore and we have to head south again, winding towards the west coast this time. In the town of Ullapool, I slip into a restaurant in search of a toilet. As I open the door to the ladies, I'm struck by a jolt like a physical blow, and it almost makes me cry out in shock. There's a huge CFL dangling naked from the ceiling, and I don't resist the urge to flee. I'd almost forgotten about them on this trip, but this experience sucks the joy from the final night of our campervan dream. Something in me sinks with every mile south.

Back home to July, I try to reconnect with friends and explain why I've been lying low. Some get it, others don't. Would I, if this wasn't me?

"Shit!" says Kirsty, processing it, pouring me a large glass of wine. "That's big. Those lights are, like, everywhere. What about shops . . . train stations? Petrol stations?" As we drink, more light encounters occur to us: "Oh, what about the doctors? And the school? The library? What if the kids need to be in hospital?" I feel a rush of gratitude.

Incandescent

Carolyn swiftly disposes of any CFLs in her home, telling me she didn't like them anyway. She replaces them with halogens throughout the house: "You need to be able to call round anytime," she says.

I find myself creeping around friends' houses, peering up at light fittings, switching lamps off. Some ignore me and switch them back on; others are bemused; many are affronted. As a general rule among my friends, the greener the politics, the grimmer the lighting. And they are proud of their low-energy lighting – some have gone to considerable expense to convert their homes. There's something irksome, threatening even, about my messing with the clarity of Bad old lighting and Good new energy-saving bulbs.

It doesn't help that my head is spinning with snatched fragments from my Googling and my growing suspicion that there is nothing "wrong" with me at all, but instead something toxic about these light bulbs. My explanations come out as forceful blurts, spiked with random rants about mercury, macular degeneration and skin cancer.

I know how I sound. How easy it is to think that something harmful to you must be intrinsically bad for everyone? I have no shortage of people in my life who feel that way about wheat, gluten, sugar, dairy products, computers, cats . . . I know well enough how it feels to be on the receiving end of such unsolicited advice. How weary are we all of being told things are bad for us? Cancer stalks my generation like a hungry raptor circling overhead. We live with one eye keeping a wary glance on the sky, wondering whom it will pick off next. And our Facebook feeds fill with regular warnings about how we might be making ourselves more susceptible to its talons . . . because of food choices, exposure to Wi-Fi or sitting in the sunshine. And now here's me adding light bulbs to that list.

And yet . . . I call round to my neighbour's house and glance in her window before going in. She's in her armchair engrossed in a book, a bare CFL in the reading lamp just inches above her head. So close. I have to stop myself from calling out "Get away from that thing". I want to spout the government guidelines at her: "CFLs should not be used in close proximity (distances of less than 30 cm or one foot) to people for longer than one hour." I go inside and she switches off the lamp for me. I settle beside the wood-burning stove and we chat awhile, until the room recedes and my vocabulary fails me. Damn it, there's another CFL in the lamp on the bookcase and I hadn't spotted it.

That night I dream I'm being fed the contents of a paper shredder. It's so dry and I'm so thirsty, struggling more with each wrinkled thread, gagging as they catch and accumulate in my throat. I wake and my mind is racing: What are those government guidelines based on? Do these bulbs increase the risk of skin cancer? If so, why are they on the market? How can light even make me thirsty?

•

By August, the green of the woods is deep and rich. Light spills in rippling pools on the woodland floor. These woods don't look like much on the map – just a thin ribbon of green around the fields and down to the canal – but they are bursting with life and tree-filtered light. The upside of all this light obsession, I realise as I walk, is an intense appreciation of the beauty of natural light. I've always noticed it, but now I revel in it. Here in the woods I gaze at the shifting spotlights, the holly sheen and birch-bark gleam; find myself full of wonder at the miracle of photo-synthesis, through which light powers all of life on Earth. And at home I find myself stopping and staring, tea towel in hand, transfixed by sunlight falling on a white wall, on wood, on skin.

Does light affect us more than we realise? I search on. Yes, I find, it affects almost everything – our mood, sex drive, hormones, metabolism, energy levels, skin and more. I've been Googling too long, and my daughter's been watching TV all this time and needs a bath before bed. I slip in with her and we splash and sing and play with plas-tic boats. I'm aware of the light shining on the silver of the taps, the painted dot of light in every rainbow bubble of our bath. August evening light falls on our skin: the gloss of wet shins, the matt of dry knees. Light rests gently on the white tiles and warm wood. The dog barges in to say hello and leans his chin on the side of the bath, pushing his face towards mine until my eye is close to the gleam of light on his wet black nose.

That night I fall asleep with my mind full of peaceful images of rainbow bubbles and girly giggles. Then wake suddenly at 4am again, the questions swirling: By what mechanism did the EU manage to ban incandescent bulbs? Who is making money out of this? How could this happen in the US where even guns haven't been banned? It carries on and on, until the morning light.

We've all been invited to a party. I check the weather forecast: it's going to be sunny so it should be okay to spill out into the garden – no need to worry about lights in the house. We'll go for a while anyway then slip away, Cinderella-like, before night falls. September's nipping the evenings tighter, bringing us closer to the winter darkness, but it compensates with a rich, golden light that holds me in the present moment, sweetening the apples in our friends' garden and casting warm tones on the hills. One of my oldest friends is at the party – I haven't seen him since my light troubles began, though he's heard my story from mutual friends.

"Hello, err, how's work?" he asks vaguely, not meeting my eye. Adam's an environmental activist and a proponent of energy-efficient lighting. I can feel his hackles rising as I say life's a bit difficult just now. I've sensed this before. It's not just the yawn of hearing someone prattle on about health troubles – this issue really makes people bristle. I've crossed a line, not just by having problems with supposedly low-energy lighting, but by questioning the consensus that it's A Good Thing.

I feel it most – that indignation – among fellow Quakers, earnestly trying to live faith in action, to see and to take responsibility for the threads that lead from our daily lives to the wider world. Incandescent bulbs were quickly banished from Meeting Houses throughout the land and CFLs installed everywhere. At least I can still worship, as we meet on Sunday mornings in a community centre with plentiful windows. But my wider Quaker life feels like it has come to an abrupt stop, including my visits to Woodbrooke.

I try to meet people outdoors as much as possible. One evening I walk with my friend Jessica up a nearby hill that's soaked in September sunshine. I tell her my story and she listens intently. It's like light has the kind of significance in your life that it used to have for people in the old days, she says. Jessica remembers visiting her grandparents in rural Italy, where they lived without electricity: darkness falling and dawn arriving therefore felt incredibly important. When she later studied poetry, her teacher described the significance of sunset and sunrise in poetry before electric lighting – these times shaped each day.

•

As September cedes to October, we follow our family tradition of an annual pilgrimage to Aberlady Bay to watch thousands of pink-footed geese pour into the estuary from East Lothian's expansive, light-washed skies. We scan the darkening sky, blushed pink and orange, and see nothing at first. Then just a shimmer in the distance faraway, a fluttering line draped and dragged through the sky. It comes closer, darkening and thickening now, with more lines behind it like the waves of an approaching army. Closer still we hear the raucous chatter of their incessant calls. "They sound like they're going to a party!" says my son. We watch them come and come, in long, flowing strings, and they descend into the shining bay until it's almost too dark to see them, but we can still hear their cries.

This is it – the tipping point of the year. I can feel the air sharpen as the seasons turn. Halloween is carnival time around here. Spider webs dangle from the gutters, gardens are full of ghosts and ghouls and scattered body parts, and lights shine from pumpkins on the doorsteps. It's one of the highlights of my son's year, and I want my daughter to share the fun in her little witch's dress, a plastic bat flapping from the rim of her hat. I pull a witch's hat over my own head and dash past the fluorescent street lights on our road to a nearby estate, which has warmer, orange ones. As we approach a skeleton-clad door, an automatic light clicks on with a shudder; it's astonishingly bright and harsh and I cower in the shadows as the children sing their Halloween song. I shut my eyes. I'm sure I can feel the light throbbing. When did outside lights get *this* bright?

That night I have the worst dream yet. I'm on the ground, unable to get up, and there's dog shit in my mouth. I wake, gagging, at 4am yet again. Grim as they are, I'm almost admiring the ingenuity of my subconscious in coming up with these vomiting dreams.

November is the month I've been dreading all year. The last of the leaves dangle like forlorn decorations outside my bedroom window, fewer of them each day. I'm aware of the colours in the woods softening, gentle and subdued, that soft, sad yellow. I don't drive in the dark anymore so my husband has to get home in time for all the after-school duties – swimming trips, taekwondo, parents' evenings, shopping. What would I do otherwise? The sense of dependency makes me feel grateful and vulnerable and resentful all at once.

This is the time in which so many indoor workers in Scotland leave home in the dark and return in the dark, barely stepping outside in daylight unless they snatch it in their lunch break. Thank God I work from home and can walk the dog and savour what daylight there is each day. Still, I feel the loss each day as the light fades. As it slips away I feel a tugging emptiness akin to grief. Log fires and candlelight seem to help most; they soothe and uplift, re-set something in me.

One December evening, BBC Four's *Hidden Killers* comes on the telly. It's gruesomely gripping. This episode, "The Edwardian Home", tells of the technological era at the turn of the twentieth century, which brought a surge of new inventions and a fervour for "untried and untested mod cons". The most exciting was electricity – "clean, invisible, magical energy" – and its first domestic application: the incandescent light bulb. A bare cable hung in the centre of the room, with multiple wires cascading from the light fitting to power other equipment such as irons, vacuum cleaners or kettles. The resulting overload was a common cause of fires, but then along came a "wonder material" – an insulating, non-flammable, resilient mineral called asbestos. It was so versatile, strong and cheap that it was used throughout the home – in electric heaters, lining water tanks, in floor and ceiling tiles. The alarm was first raised by a factory inspector in 1898, and post-mortems of workers showed lung damage, but these findings were ignored, "partly because of cover-ups, partly because of a desire not to know". The first formal diagnosis of asbestosis was not until 1924, and legislation took much longer. "Sometimes it was ignorance," says one commentator, "other times it was profit – there was so much money to make out of it."

As I stroll back through Edinburgh after a work meeting on a December afternoon, I can't help but notice that Christmas lights are everywhere, brasher, brighter, harsher than they used to be. It's not just the colours – it's the soul of the lights that has changed. As if there's a little less magic, less Christmas spirit in them now.

The main extravaganza hangs over the glamorous shops and fancy restaurants of George Street, where there are thousands upon thousands of light bulbs. The street below is packed full of people looking up to admire them, but I rush through the crowd to get to the station. I feel worse on some trains than others, guessing it's the age of the stock as the older, yellow fluorescent lights don't affect me so much. I slump into the seat on my train grumbling inside: if the incandescent ban is really all about saving energy, could we not tone down all the lights a wee bit at Christmas? How come councils throughout the land can fill the streets with a thousand lights each December night, but we're not allowed one incandescent light bulb for a bedside lamp? It just doesn't make sense.

Our local church is a hub of festive warmth at this time of year and we've walked across the snowy park many a

Christmas Eve to coloured lights and candles at the family service. I want the children to experience the church on this special night, but there are CFLs in there too, and I don't want to fill my Christmas night with vomit dreams. I wait outside, pacing around the car park, trying to join in the carols through a lump in my throat.

The light comes so late on January mornings – almost grudgingly, it seems, as we leave the house for school. Then it's snatched away again almost as soon as we return in the afternoon. There are dull days where it feels barely light at all, but on bright days there are moments of incredible clarity, when I can pick out astonishing details in the hills and houses and castle, and when even more mountains appear behind the familiar outlines. Everything is sharper, as if brought into focus by the brilliant winter light.

Blue Monday. The third Monday in January is supposed to be the most depressing day of the year: festive hangovers still hitting body and purse, and the darkness of winter stretching ahead. But I open my door and gasp: the hills are shining with pink, dawn-blushed snow. As darkness falls, I stay close to the fire, holding tight to the thought of summer, the stack of firewood, the friends-who-changed-their-light-bulbs close by. I read the work of a Finnish writer friend as a lesson in winter survival. The Nordic peoples have an acute perspective on light and darkness:

> You must understand that in Finland, time is not bound
> to a clock. Time is dependent on light. The winter days

are awfully dark and we just have to trust that there will
be light again some day.

One morning in February I wake up at the usual time,
a bit before 8am. Something has changed: there's a tangi-
bly different quality to the morning. I yawn and wriggle,
coming round, trying to sense what it is. Hope? Relief? Oh,
it's light! I open the curtains and see, behind bare branches,
hints of yellow and blue in the grey-tinged light of the sky.
Morning light! Welcome back, I can't tell you how much
I've missed you.

And now it's got a hold on the mornings, light picks
up surprisingly fast. It's a steady crescendo of hope rising
through the month: each morning a touch lighter, each
evening reaching a little bit further. I feel energy flowing
again, a sense of moving forward in productive, creative
days and passionate nights.

•

It's time for my appointment with the dermatology
department at Dundee University Hospital. I'm not overly
hopeful.

I've already seen one dermatologist who told me: "No,
there's nothing new about these lights; neon lights have
been around for decades." I'm not talking about neon
lights, I tried to explain, but didn't get through. I've also
seen a neurologist (at a hospital wholly lit by bare CFLs),
who was kind but proclaimed me healthy and could offer
no explanation. Still I keep seeking second opinions, so that
at least I can keep asking questions.

Incandescent

So now I'm in a hospital gown with an open back, and a friendly nurse is drawing circles on my skin in biro. Different frequencies of light are shone onto the circles on my back to test if there's any reaction. There isn't.

Do a lot of people have trouble with CFLs? I ask the nurse. Oh yes, she says, they get enquiries from all over the world. The people here *are* interested: a panel of six specialists question me. I tell them it feels like it's affecting my brain not my skin – can they tell me anything about CFLs and the weird feeling in my head? No, this is a dermatology department. They suggest getting a second opinion from another neurologist, but I'll have to go back to my GP for a referral.

Half-term holidays take us north to Pitlochry, leaving too late – as ever – and arriving at our holiday cottage in the dark. It's a Christmassy thrill to anticipate morning in a new place, not knowing what the light will reveal: I can hear water behind the house and can just make out the dark bulk of trees against the slope of the hills. I let the dog out for a wee and gasp. The stars! The sky is thick with them, each one polished to brilliance. I've not seen stars like this for years. I call my husband and my son emerges too. He stares, open-mouthed: "They're twinkling," he announces. "They really do twinkle!"

When I wake, our trees and glen are still in shade but the hill behind is catching the low sun and gleaming. The far fields are all striped with golden snow and blue shadows. The day is picture-postcard pretty in precise February sunshine: the river rushing through banks of snowdrops, and the mountains around the town shining, white against the blue sky. That night I walk the dog for over an hour through the steep, snowy streets, the snow glowing orange under the sodium street lights, warm and familiar. I walk and walk, until I can almost believe I'm safe in a former time, before this lightmare ever started.

Back home, I'm sitting in a chair by the window thinking about light when a rainbow appears over the Forth and across the big sky. There is a yellow note to the light now, warm and welcoming. Hello, March. It's 5.30pm – a few weeks ago I had the curtains tugged shut against the fluorescent street lights by this time. Now I'm looking at the soft light sweeping in from the west and filling the clouds. I watch while it plays with the Ochil Hills like an indecisive artist: picking out glens and rocks in sharp relief with strong brushstrokes, before changing its mind and smudging the whole scene in a broad pastel blend.

But that night the questions surge and jostle inside my head with such ferocity that I writhe in my bed. I reach for my notebook and start to write whatever's in my head, all the rants and queries, just getting it down. And as I write, I begin to relax until, finally, I can drop my pen and sink into sleep.

•

On a bright April morning I'm heading off to look for otters on a riverbank. I'm still writing about wildlife for a living, and working closely with photographers has helped to tune me in to light. I've noticed how alert they are to where shadows fall, to colours changing subtly throughout the day, and to how the oblique angle of morning and evening sunlight creates a richer, warmer tone – "sweet light", they call it.

It's work I love, but the difficult part is the getting there: two train journeys means three stations, and lights to negotiate everywhere, from the trains themselves to the ticket office and toilets. And the anxiety about not knowing what light I'll find where augments my claustrophobia so that as the train doors shudder and clunk closed, a heavy claw grips my shoulder blades and another cramps around my lungs. *What if*, what if they don't open?

I see myself – dark-eyed and worried-looking – in the train window as we pull out of Waverley station. This isn't me, I'm thinking. Where's the me who used to travel so freely? Who would think nothing of popping to London or Dublin just to hear a band play, untroubled by squashed or missed or late trains, always with a book and a daydream. Now I'm shaking all over because I'm leaving the familiarity of Edinburgh, where I know which toilet has a halogen light bulb. When did my world get so small and scary? I feel that life is closing in as new lighting spreads.

But breathe: the lights are fine on this train, and soon I'm getting off. And the doors do open. We spend the day walking by the river, watching the water. There are tantalising signs of the presence of otters: tracks on the soft mud by the water's edge; silver smudges of fish scales on the path. There! The photographer motions with his eyes to the water. It's not an otter, but a flash of kingfisher, bright and brilliant. We settle a while on the riverbank, the water still bright even as the sky around us begins to darken. I'm still hoping to glimpse an otter at any moment, and the silver rings of fish rising make me start, as does a sudden movement on the water's edge, but it's just the rippling wake of a mallard pootling along upstream.

I realise I'm grinning from ear to ear. My photographer friend gives me a quizzical glance, but doesn't ask. I couldn't have explained anyway. I just wish I had a way to keep this moment – the sweet smell of the air and the evening light lying like a gilded skin on the river's surface. I want to take it and use it as a shield to hold up against the monsters of the night. Look, I've still got this, I want to say. As long as I can still do this, I'm going to be okay.

2

Other People's Stories

I begin to blog. At first, it's just a way of working through my own concern and confusion. This issue gets ever more complicated the more I learn about it. Questions only leading me on to further questions, and writing is my way of processing stuff and gaining some kind of clarity. And I'm weary of trying to explain my situation. Some of my friends can't understand why I find going out so difficult. Even meeting for a pint is complicated – I have to recce the venue fist, try to find out what lights are where. You can't phone and ask as lighting involves a vocabulary that few understand. It is hard to explain all this to friends and family without opening the floodgates. So I stick my thoughts online – it's somewhere to put them and to direct people to if they ask.

I don't anticipate what happens next. That people I don't know somehow find my blog and respond by sharing their stories with me:

The city has been installing LED street lights in our area. The one road I drive for five minutes to get home at night has become covered with LEDs. Last night after work I drove down that road and searing pain went deep into my head. My mind felt like it went into shock. I felt like I was in a microwave oven. I felt fried. The pain actually increased over the subsequent hours. Now, twenty-four hours later, it is subsiding, but it feels like the spasm could be re-triggered more easily. Even as I am looking at my computer screen, the light is bothering that part that feels wounded – my computer doesn't normally bother me. OMG . . . I don't know how I am going to get home from work.

I have had problems with fluorescent lights and low-energy bulbs for years now. I collapse all the time. Collapsed at the doctors last Tuesday and was sent to hospital. They can't find anything wrong with me. I need to see the doctor regularly for other problems but now I am scared to go. My eyes are always very sore and I have blurred vision. The doctors don't understand how bad it is. I tried to end it all two years ago but failed. This has had a huge impact on married life. We can't go to many buildings like normal people. I can't even go to public toilets. This has totally ruined the life of myself and the life of my wife.

The stories keep coming. I hear from teachers and lecturers who had to leave their jobs because of new lighting; students who couldn't continue their studies; a single mum who couldn't go to her daughter's school shows; a barrister who had to leave the country because the new lights in the law courts "feel like someone is clanking my head with a hammer"; a lady who tried seven churches, despairing of finding somewhere to worship on a Sunday.

Hearing these stories from around the world makes me grasp that there is a much wider context here. And the more I hear, the more it bolsters my growing conviction that this is *real*. This isn't some psychological antipathy to new lighting, as people often suggest, but actual physical reactions, which suggests there is something wrong with the new lighting technology itself. Realisation comes to me as if in sudden bullet points:

- This isn't just me.
- My symptoms are relatively mild – there are people out there having a much worse time than me.
- This is way bigger than CFLs. All kinds of new lights are causing suffering, LEDs especially.
- This is physiologically complex – different lights are having diverse effects on different people. And the same lights are having different effects on different people.

There is a disparate global community of people meeting online and asking the same questions. Some have existing health issues that are exacerbated by new forms of lighting. Most – like me – are in good health except when exposed to the lights. We share and compare experiences, enjoy the relief of some dark humour and of light-related jokes. It seems every person adversely affected by new lighting has to go through the same process of trying to understand what is happening to them and to communicate that to others. I keep hearing a familiar story: the people who should hear us in times of distress – the church leaders, the unions, the political representatives – just don't want to know because they are "doing their bit for the environment".

It's as if we are just an inconvenience, collateral damage from an EU policy on energy efficiency. But these are real people. As we get to know each other, over emails and forums and Facebook chats, some of them begin to feel more real and closer to me than those long-standing friends who do not – cannot – understand. And one particular question keeps arising: is there something "wrong" with us light-sensitive people, or are we simply more "sensitive", with a heightened ability to perceive some aspects of light?

It is like living in two worlds. In my day-to-day life – going to the park, picking the kids up from school – most people have never heard of anyone having health troubles from lighting. They look at me like I have two heads if I say a particular light bulb makes me ill. And yet, somewhere below the surface in the digital world, people all over the planet are Googling words that eventually lead them to each other and to my blog. It feels as if something is happening, the way fungi spread and grow along an underground network, forming mycorrhizal relationships with plant roots to share nutrients. People are reaching out to each other, forming links, sharing information, sprouting blogs. I have the sense of something ripening and ready to burst to the surface.

Incandescent

I'd like to introduce some of these real people and share their stories . . .

Elaine (Ireland) • Elaine always lived an independent life with a love of the outdoors, animals and nature; a young, active woman living in a small town in the West of Ireland. She'd always been a night owl, and particularly loved being outdoors after dark – heading out for a run after a day's work. She also loved the magic of early morning, and would often be out walking the dogs before sunrise. When she was twenty-eight years old, Elaine decided to go to Australia and bought a Nintendo game for the long journey. As soon as she switched it on, she felt a shooting pain behind her eyes, so immediate and intense that it left her reeling.

The Nintendo was backlit with LED light. It was Elaine's first indication of an extreme intolerance to this form of lighting, which would come to devastate her life. A game can, of course, be avoided, and so she immediately abandoned it and was able to enjoy her time exploring Australia. But she returned home to the beginning of a new era of lighting. Like an enemy army approaching and encircling her, LED light began to encroach on every sphere of life, and to take it from her, piece by piece. "LED seems to have taken over the world," she says.

The first blow was her work, when new computers with LED backlit monitors were installed. "I was already struggling and extremely sick at work with the overhead LED lighting. I was living on painkillers and anti-sickness tablets just to get through the day. I couldn't tolerate the monitors even for a few seconds so I had no choice but to leave my job."

Next came the installation of bright LED street lights right outside Elaine's home.

"When the new street lights were erected I couldn't even step into the front garden without eye pain and pressure, migraine, dizziness and nausea. The pain is similar to when you fall over and bang your head on concrete: that nauseous feeling you get just after the impact. The symptoms start after only a few seconds' exposure, and the pain continues to get worse for hours afterwards, even after going back indoors."

The street lights arrived in winter, and imprisoned her in her own home from the moment they came on in the late afternoon until after 8am the next morning. Elaine could no longer walk the dogs in the evening or early morning – if she got the timing wrong, perhaps by stopping to chat with a neighbour, she risked getting "caught" by the lights coming on, which would mean an evening of pain and illness ahead. Even inside the house, Elaine could sense the lights when they came on: "I hear a ringing and my ears hurt and get blocked." Again Elaine felt she had no choice but to leave her home and move away to the countryside.

Still there was no escape: bright LED lighting was soon installed outdoors on neighbouring farms and houses, meaning she could only access certain parts of her garden; and the increased use of LEDs in car headlights, as well as the rise of daytime running lights, made driving – or even walking – difficult at any time of the day. Christmas made life even harder: Elaine has to spend almost a month indoors at night to avoid nearby displays of LED Christmas lights.

"When the Christmas lights are on there is a sense of being trapped and isolated, like I'm locked in the house – it's a feeling of claustrophobia or panic that I never felt before. I know I can't go outside with my dogs because the symptoms are too severe. It feels as though my freedom and independence are gone and those things are really important to me. It's very distressing and it really takes a toll over the month. I also think about the future: what if I have to do this for the next fifty years?"

Having fled the town for the hope of sanctuary in the countryside, Elaine feels back at square one again. "I don't know where else to go," she says. "A fairly isolated life is the lesser of the two evils, but no one should have to live in isolation like this because of a light source. I hope and pray this gets resolved in our lifetime. I'm in my thirties, but I often feel I am living the life of an old lady." She hasn't been able to work since leaving her job and can see no prospect of doing so while LEDs continue to affect her so badly.

Elaine has tried various glasses and lenses, including "blue-blockers", and seen all sorts of medical specialists, but as yet no one has been able to explain the severity of her symptoms or offer anything to mitigate them. "I'm nearly certain I am affected more neurologically and that it's possibly something to do with flicker and directional intensity. The older forms of light have a softer, dispersed glow," she says. "The first neurologist I went to a number of years ago more or less brushed me off, he said he couldn't help me because LED technology is too new." She now visits two neurologists (which is expensive) and is trying various treatments – but they have their own negative side effects and can be exhausting.

An ophthalmologist recently confirmed that Elaine's eye health is good, and she is otherwise fit and healthy. "I feel it confirms that there is something very wrong with this form of lighting. I don't understand how I can be young and healthy but in bits after just seconds of exposure to LEDs. Yet after a few days away from them, I start to feel more normal again. So it can't be me. It *must* be the light."

Elaine is now hearing from many more people with LED sensitivity, in Ireland and around the world, and thinks that problems with light are more common than is generally acknowledged.

"A clinical optician told me he is seeing an increase in people presenting with my symptoms, and has been for some time. Maybe my level of sensitivity isn't as rare as I once thought – I do think there are many others suffering in silence, even those who have milder symptoms. No one should have to put up with headaches in their work environment, or experience pain just to go the shops, the gym, the hospital, friends' houses, the church. But people do put up with it, because they believe it doesn't matter – they are just one person. However, it does matter. Every person matters."

Incandescent

Jesse (USA) • Life was going well for Jesse back in 2013. He'd just been promoted in a job he enjoyed, he had an active social and sporting life, and he was engaged to be married to the love of his life. There had been a few health issues – back injuries from snowboarding and bodysurfing adventures – and he had experienced some headaches and eye strain when the fluorescent lighting at his work was upgraded, but he found he coped okay with anti-glare glasses. It was not enough to get in the way of the sense that he had everything going for him.

Just after he married, Jesse's workplace refurbished the lighting again, changing from fluorescent lighting to very bright LED strips. "There is no description for the pain these lights caused me," he says. "It just destroyed me for weeks as I was in a solid state of migraine. I would recover a bit by the weekend but then be straight back into it on Monday."

Because he loved his job, and the life it facilitated, Jesse hung on as long as he possibly could. Too long, he realises in retrospect. The company did what they could, moving him around to different areas of the building and experimenting with adjustments. "They tried their best," he acknowledges, "but at some stage, HR make decisions for you and in the end I got obsoleted – it was the beginning of an extreme adventure with lighting."

There followed "a hellish period of about three years, seeing various neurobiologists and taking strong medication, which did nothing to help and some had scary side effects. The medication put me in a much worse place."

He tried hard to carry on with a normal life – shopping and travelling and constantly pushing himself to be on the computer. But it just made things worse. He learnt the hard way that actually he had a shrinking window of tolerance to LED lighting, and that in fact the effect of exposure was cumulative.

"It's like a bucket that can only take so much," he explains. "And when that bucket overflows you're in a terrible state. It doesn't go away if I leave the environment – at times it can take a month to recover. I really thought it might all be in my head for a while – I thought I'd get used to it – but at some point I realised I was adding to that bucket every day."

He learnt to pay a lot of attention to how full that bucket was – the longer he could go without any LED exposure, the better he would feel. "I think the brain heals, but it takes time. I do wonder: if I could go two years with no LED exposure at all, could I get better?"

Jesse's doctors and neurologists are trying to find out what they can, and are investigating whether nerve damage from previous back injuries could relate to his response to lights. "I definitely feel they're on my side. And I'm lucky, a lot of people with lighting problems just get told that it's 'psychological issues'. I think that's really harmful as the more exposure you experience, the worse you get – I'm absolutely sure that it was the period of intense exposure at the end of my job that made my sensitivity so extreme."

Jesse feels that both flicker and blue frequency contribute to his reactions to LED lighting. "Flicker is definitely an issue for me. I believe sunlight is the healthiest form of light and it helps me to heal, but I've noticed that even sunlight flickering through trees or a fence can cause symptoms – not as aggressive as with LEDs, but it shows that flicker is an issue for me even with full spectrum light. I seem to have a flicker threshold that means the faster the flicker, the more severe the symptoms. A lot of people who have migraine issues with fluorescent lights find that LEDs are okay, but I've come to understand why fluorescent light is not as bad as LED for me: LED has a very extreme 'on and off' flicker, whereas fluorescent light is a gas that glows so it always leaves a ghost light, and it's never as abrupt.

"A very bright blue light can bother me almost as much as LED – even an incandescent bulb when inside a really white fitting, so white that it's almost blue. And maybe this is part of the problem? LED lights in any colour are technically a blue light filtered through something – a plastic cover then phosphorous coating – to change the colour."

Jesse wonders sometimes whether worrying about lighting makes the problem worse, but he recalls one early incident that convinced him this was not a psychological issue. He was out for dinner with his wife at a favourite restaurant, relaxed in the knowledge from previous visits that it used incandescent lighting. "We'd been there about half an hour when my wife said that I'd stopped making sense. I can get brain fog and really bad word dysfunction, with words going backwards and forwards and I can't speak properly, as if I'm having a stroke. I was pretty far gone at that point and had a migraine within the hour." It turned out that the restaurant had replaced some of the incandescent light bulbs with LED ones that look similar.

Jesse's wife takes a protective role, looking out for warning signs such as slurred speech or a sudden rise in body temperature. She's constantly on guard to stop other people showing him their mobile phones – even friends who know about his problem but still do it out of habit. "It's become such a normal part of life to pull out your phone and say, 'Hey, take a look at this.' People forget . . ."

The arrival of a beautiful baby boy brought great joy but also new challenges to Jesse's light-sensitive life. The family moved to the countryside, close to grandparents and with five acres of land so that the neighbours' lights are not within sight. Jesse's wife goes out to work while he stays home with the baby, "feeding him, changing nappies, trying to make a contribution that way". He can take his son to parks in a nearby town during the daytime, but he doesn't go out at night.

Jesse doesn't see any prospect of returning to the workplace unless his light sensitivity can be resolved. This results in some guilt and a lot of anxiety about life ahead with his son. "It's hard – he's going to have a dad who can't do everything mum can do. It makes me sad, and I understand why people around the world are protesting about LED street lighting. They're cold – it's a deathly look! In addition to all the other issues, I can't see well under LED. Everything looks grainy and I have a hard time making out shapes properly.

"I could get really mad about it, and I went through

phases where I might have been angry, but I see it in context. Look at asbestos, and many chemicals – the human race is plagued with things that make individuals sick, but until there are too many people affected, things don't change. You have to hope that technology gets better. LED lighting is changing and progressing so fast, but I don't yet know if I'll ever be able to tolerate it. I haven't found an LED bulb yet that's okay for me, and I've tried quite a few. You get tired of experimenting on yourself – it's kind of hard to put your hand on the hot stove over and over again."

Joan (UK) • Joan still remembers the first day she walked into a science lab. She was an eleven-year-old schoolgirl. "I thought it smelled wonderful. It *was* wonderful," she recalls. That delight never left her and eventually led to a thirty-year career as a lecturer in pharmacology, which she absolutely loved.

Ever since she was a teenager, Joan had suffered occasional migraines, which were hormone-related and always took the same form: loss of vision and disorientation but no pain. Much later she began experiencing a different type of migraine, with intense and increasing pain and loss of balance. These migraines seemed to be triggered in some way by indoor lighting. Things then got worse when one part of the building where she worked was refurbished: she took one step inside and immediately collapsed onto the floor. "My balance just went. I fell over as if I'd walked into something."

What she had walked into was the "staggering brightness" from the new type of lighting on the ceiling. Realising she could not work in this environment, she appealed to her trade union for support and to the university for her seminars to be timetabled away from the new part of the building. Although the university acknowledged that Joan did have a disability as defined by the Equality Act, Joan ended up having to go an employment tribunal, claiming that the university would not make "reasonable adjustments" to her working conditions.

Joan lost the case. The tribunal ruled that "allowing the claimant [Joan] to carry out her teaching in buildings other than that part of the building whose lighting causes the migraines would not be a step that it is reasonable for the respondent [the university] to have to take to reduce the risk of the claimant's migraines". Its reason for the judgment was as follows: "While the students would have the benefit of the newly refurbished buildings for other lessons, we recognise that they may feel let down if they are denied the use of them when being taught by the claimant. They may also feel disadvantaged if they are taught in poor light. We can also see that if students feel dissatisfied, this may have an impact on the ability of the respondent to attract students in the future and it could become a threat to their commercial viability".

The ruling was reported in *Private Eye*, which picked up on the irony of the presumption that students would value a particular venue over an engaging and popular lecturer:

"There is no mention of how let down the students would feel at losing a senior lecturer with twenty-seven years' experience in the field." The ruling also presumed that the old type of lighting is in some way inadequate.

The trade union's lawyer, who had represented Joan, advised against an appeal, saying the case could not be won, and so Joan reluctantly took early retirement, leaving the job and working environment that she had loved. She took up distance and online teaching at a fraction of her former salary, and without the aspects of the role she had found most stimulating and rewarding: the direct student contact, lecturing, research and travel.

Around this time, Joan was elected as a local councillor. This presented a logistical challenge as some rooms in the council's chambers had lighting she couldn't tolerate. The council tried to accommodate her, even erecting a tent in one of the rooms to hide the lights, but that didn't work and so she tried to join in meetings from home, via the internet. "I could express opinions, and good chairs ensured I had a chance to do so, but I couldn't totally fulfil the commitment to my constituents as I couldn't vote."

This is because the law in the UK requires councillors to be physically present in the room to vote. Council officers did their best to find loopholes: are exceptions made for councillors in the Highlands and Islands of Scotland or other areas where they are geographically disparate? No, these councils overcome their difficulties simply by meeting less often but for longer: the members still have to be in the room to vote. And so, although an extremely active councillor, this inability to vote may be why Joan was ultimately deselected by her political party.

As new forms of lighting spread throughout the public, shared environment, Joan had a sense of her world shutting down . . . leaving a void into which depression and anxiety crept, and spiralled. She had lost not only her job, but access to the vibrant arts and social life she had previously enjoyed. "I used to be a bit of a culture buff, spending hours in museums and galleries," she says. "As things became increasingly restricted, I knew my forms of entertainment would have to change, but to what? I thought about doing some Tai Chi, but lo and behold, there are crazy lights in the hall. I used to do some voluntary work reading on talking newspapers for the blind and partially sighted. It was a social event as far as I was concerned and the other volunteers were good company – then they put CFL bulbs in the office . . ."

It's now several years since Joan left her job, and she has felt increasingly isolated and desperate during that time. She doesn't know why certain forms of lighting have such a bad effect on her. "I'd like to know more – otherwise I sound like a crank, with people thinking I am claiming to be allergic to the modern world. I'm not."

Joan has tried to be proactive in raising concerns about new lighting, and about the EU directive that bans the importation and manufacture of incandescent light bulbs. "I went to visit one of my local MEPs to explain the problems new lights are causing some people. The MEP said,

'We have to have the directive because if we don't then millions of Bangladeshis will drown.' She actually said that! And outside there was a taxi with its engine running, waiting to take her to the airport to fly to Brussels.

"There are not a lot of parliamentarians who are scientists," she adds ruefully. "Politicians in general are not particularly numerate or scientifically trained. They have brought in the directive because it appears to be 'green', but without properly considering whether it will achieve its intended aim. What about the true environmental and energy costs of making and installing the new lighting?

"It can take a long time for an understanding of triggers of adverse health effects to lead to change. My dad was a seaman and then a shipyard worker who died of mesothelioma, which is the cancer that results from asbestos exposure. Of course, asbestos was once thought to be no threat to health. I believe this imposed lighting change is, in effect, a global experiment, and we are the guinea pigs."

Joan keeps her own home "chock-a-block" with incandescent light bulbs, but the difficulties of leaving the house means she feels she has nothing to look forward to: "I've discovered what it is to be disabled – to be asking for help and apologising the whole time. It's uncomfortable. I used to live a happy, normal, very independent life, but now I feel I don't have a life. I pretend to, sometimes, but I don't."

•

Hearing these stories, something shifts within me: I move from bewilderment to a searing sense of injustice. A fiery kind of anger begins to burn, and with it comes clarity – I catch a glimpse of the horizon and see the path ahead. It looks like an intriguing, confusing and exhausting route, but I have a sense that I can't escape it. Not just because I'm one of those physiologically affected by some forms of light, but because of a gut-felt sense of urgency to say something about it. Some Christians talk of a "calling", and Quakers recognise it as a "leading" – a compulsion to action, to speak the truth as they experience it. For me, it is like a shove forward onto a stage and a command to speak, however bewildered or reluctant I feel about doing so.

The stories begin to coalesce and I see a bigger picture emerging. Not all of it – there are questions of physics, physiology and legislation that I struggle to understand. But I see enough to grasp that something somewhere has gone very badly wrong. People are suffering intense pain and getting shut out of their own lives. And it seems nobody wants to know because they are trying to be Good and Green.

I feel tasked with saying that we need to talk about light. All of us. We need a conversation, publicly and privately, nationally and internationally. It's too big and too crucial to be changed so radically without real understanding of the consequences. Vast decisions about artificial light are currently being made by politicians and civil servants on the basis of perceived carbon emissions alone. I'm not convinced that those making the laws about the quality of light understand enough about light to do so. We need the rest of the stakeholders in on the conversation. We need physicists to explain a thing or two. We need doctors, dermatologists, ophthalmologists, neurologists and psychologists to pool their professional expertise on light and human health. We need to hear from prison officers, educators, health workers – all those who may have noticed how light affects the people they work with. We need to hear from the artists, photographers and designers who understand intuitively the effect of light on the soul. And we need to hear from biologists and ecologists about how artificial light is changing the natural world.

For each and every one of us, this is about the light in our homes and workplaces and streets. We all need to talk about light.

So, I decide, if this conversation isn't yet happening, I'd better start myself. I'm going to talk about light, with everyone I can, and see where that leads.

3

This Stuff of Physics, Metaphors and Mysteries

I am drawn to light through sensation – *feeling* it and noticing experientially its capacity to heal and to harm. And the subtlety and intricacy of light's astonishing power to uplift, torment, startle, highlight, burn, welcome, seduce . . . I could keep this list going for a long time. But I know little about equations or the physics that describe light.

There's a presumption, it seems, that light sensitivity somehow arrives pre-packaged with a working knowledge of physics. "Is it the flicker or the spectrum?" my techier friends ask me casually. "Are you okay with LED under 3000K?" "What about the CCT?" Begin to question light and you're soon embroiled in discussions about lumens per watt, colour temperatures, photons, diodes and electromagnetic radiation . . . The language of lighting has a backstory of physics, and I don't even know what some of the words mean, let alone the numbers.

I've written about science for much of my working life – it's often my job to act as the bridge between scientists and the general public. But as my specialism is nature and wildlife, I have always concentrated on biology and ecology.

Incandescent

Physics and I parted company when I was a fourteen-year-old schoolgirl, and it had been an unhappy and short-lived relationship even then. I still remember the sense of utter incomprehension, tinged with the humiliation of defeat. I thought myself and physics would never meet again. And now here I am in my forties, looking up at a conceptual mountain I feel I have to climb. My work as a journalist has given me a preliminary foothold, but I'm going to need to expand my knowledge to get to the summit.

And so I begin my journey by curling up in bed with the quilt over my head, trying to grasp the meaning of these words, which both shape and obscure the dialogue. What even *is* light? It's so difficult to get a handle on – how do you begin to understand light?

A breakthrough comes in the unlikely guise of an economics lecture. Annie Miller, recently retired from Edinburgh's Heriot-Watt University, is giving a talk at the Quaker Meeting House and speaks about the widespread phenomenon of economic illiteracy. She says that we step back from the issue, leaving economics to the "experts" because we feel we don't understand it, that it is somehow beyond our grasp. And because we fear unfamiliar language, the specialist vocabulary common to many professions inhibits us from participation.

Don't be cowed, is Annie's rallying call. Economics is far too important to be left to economists. We are all recipients of economic policy and witness its impact on people's lives, and as such we're entitled – no, more, we're duty-bound – to engage with it. This is not about numbers, she tells us, but about values and how we distribute resources. What matters is the effect of economic policy on real people: it can create comfort, prosperity, destitution and homelessness. Ultimately, it's about justice.

We listen attentively, and the sun glints on the rooftops of Edinburgh outside and gleams on the Pentland Hills beyond. My perspective shifts. Suddenly I am back on familiar territory again: human lives, values, justice . . . This is my kind of stuff. I think differently about economics – and light – from this point on. We are all recipients of light, and lighting policy and regulation. It has a profound effect on us all, on the cells in our bodies, whether we are aware of it or engage with it or not. Is a widespread lack of under-standing of physics a part of that lack of engagement for most people? Are we wary of stepping into a subject matter laced with unfathomable vocabulary, and does this result in a kind of "light illiteracy"? Ultimately, it's about how light impacts on people's lives. This, too, is about justice.

I somehow have to climb this mountain, step by step, and I start with light bulbs. Nowhere is our "light illiteracy" more evident than trying to have conversations about light bulbs. I want to book a night away in the Highlands for our wedding anniversary, but this means trying to find out what lighting is used in a hotel:

"Aye, they're just normal light bulbs, doll," says the receptionist when I phone to enquire.

Can you describe them? I ask. *Are they curly ones or pear-shaped?*

"They're just, y'know, the *normal* ones."

Incandescent

This conversation continues through a selection of Highland hotels and, despite requests to speak to facilities managers, doesn't get much further. No one can tell me if their lighting is incandescent, halogen, CFL, LED . . . It's simply not a vocabulary that many people have.

It's not surprising, really. For, once upon a time, there was such a thing as a normal light bulb. In most households, this just meant incandescent bulbs. You had a stash of them in the hall cupboard, maybe a few of different brightnesses, some clear and some pearl, but that was that – changing a light bulb was such a simple affair that a whole genre of jokes grew up around it. But in the last decade or so, changing a light bulb has got a whole lot more complicated. Visit a hardware store now and you're faced with a dizzying array of bulbs and fittings in every shape and form, and a rolling wave of technology, propelled by commerce, legislation and ideology, ensures there are more each year. What information do customers have on each of them in order to make a choice? So many homes and businesses now have a random mix of new purchases and leftover stocks. Eventually I send my husband to recce a Highland hotel. On every lamp up the wide staircase there seems to be a different bulb: a CFL here, an incandescent lamp there, an LED a few steps up the way.

Few people can speak the language of light, myself included. I can feel a difference between a CFL and a halogen, for example, but cannot really articulate it. I grasp that there are different means of creating light, and these are exploited in different types of light bulb: incandescent, halogen, CFL, LED, and so on. But what I'm really seeking to understand is whether there is any difference in the light they give out – is the light from a CFL or LED bulb the same *stuff* as light from an incandescent bulb? And how does that affect us – all of us in generic ways, but many of us in individual ways? How can the same light give one person eye pain, make another vomit, and have no (apparent) effect whatsoever on a third person?

I need to get to grips with some core principles of physics, and I'm pondering this one evening when a BBC programme on TV catches my eye: *Scotland's Einstein: James Clerk Maxwell – The Man Who Changed the World.* Glaswegian presenter Professor Iain Stewart is gently berating pedestrians in central Edinburgh for passing by the statue of James Clerk Maxwell without noticing him, or knowing who he was. "But this is Scotland's greatest scientist!" Professor Stewart exclaims. This man's work created major scientific breakthroughs in our understanding of colour, light, astronomy and electromagnetic radiation. His equations laid the foundation for the modern world with our radios and radar and microwaves. His contribution to physics was so significant that Einstein had a photograph of Maxwell in his study and said of him: "One scientific epoch ended and another began with James Clerk Maxwell."

Guilty as charged – I've spent many a lunch hour browsing the fine shops of Edinburgh's George Street without ever having been aware of the statue of Maxwell, up there on his plinth with a colour wheel in his hand and dog at his feet. But now I'm sitting up and listening, for this story is resonating now, brimming with significance.

Born in Edinburgh in 1831, Maxwell was fascinated by physics and its mathematical language from an early age – he published his first scientific paper at fourteen years old. While studying at Cambridge University, he became intrigued by electricity and engaged with the work of another scientist, Michael Faraday, who was exploring magnetism. Both electricity and magnetism had long been known about but were considered completely separate things. But Faraday was conducting experiments that showed iron filings responding to a magnet, and magnets and electricity "affecting each other through thin air" – how could that be?

Combining Faraday's research with his own work on electricity, Maxwell realised that electricity and magnetism were inextricably linked. Then his mathematical equations revealed that they were two aspects of the same phenomenon – a single electromagnetic field. He devised an equation by which this could be calculated and saw that the common parameter was the speed of light. Eureka: light itself is an electromagnetic wave. Visible light – and the rainbow of colours from red to violet that we can see – is just one sliver of this huge, continuous spectrum of electromagnetism.

Maxwell was working in an era when physics was physical, known as "natural philosophy" and all about machines and cogs and tangible objects. His work was met with a "bewildered silence" and he was thought to be "away with the fairies" because he was asking people to accept that an invisible something could affect matter from a distance. It's hard for us in the twenty-first century to understand how difficult it was to believe this then, says Professor Stewart. But is it? I'm thinking of the sniggers in the corridor ("Someone says they're allergic to *light bulbs*!"); friends' sideways glances to each other; and the raised eyebrows that greet me when I assert that I can *feel* a CFL, and I know where one is in a room because something is reaching me, affecting me, moving through the air from the lamp to my head.

In 1874, Maxwell helped establish Cambridge University's Cavendish Laboratory, where he continued his investigations into light and electromagnetism as the first Cavendish Professor of Experimental Physics. He encouraged his students to experiment with an open mind: even if you don't discover what you're looking for, you may find out something else, he said.

I get an opportunity to follow Maxwell's footsteps to Cambridge, and find myself crossing Clerk Maxwell Road on my way to the Maxwell Centre, which fronts the Cavendish Laboratory today. It continues to be a world leader in physics, counting leading scientists and Nobel Prize winners among its alumni. I'm shown to the nearby Cavendish Museum, which displays the original apparatus that Maxwell used to explore the composition of light: a simple colour wheel, red, green and blue, and a more complex light box which splits light through prisms into its component colours.

Maxwell came to understand that white light could be created using red, green and blue, and he proved this with his colour wheel and light box. This work is the precursor of all forms of modern colour display on our televisions, computers and mobile phones today. I'm intrigued to learn that he tested his experiments extensively on his Cambridge colleagues, understanding that the colours he saw were not necessarily as others saw them. He knew that our perception of colour varied between individuals, and could observe through his experiments that some people were colour-blind.

Maxwell didn't actually live to see his theories proven. It wasn't until German physicist Heinrich Hertz proved the existence of radio waves in 1886 that Maxwell's predictions about electromagnetic radiation were confirmed. But Maxwell worked it out by mathematical equations. Just as he was the first to figure out – before telescopes were powerful enough to confirm – that the rings around Saturn were comprised of solid objects: bits of rock whizzing around in orbit.

I can see a look of wonder on my face reflected in the glass case housing Maxwell's equipment. My perspective has altered: numbers – and the shapes and hieroglyphics that accompany them in equations – shift from being the components of the sums I dreaded long ago into transcendental building blocks with which you can create great pathways of understanding into space and time, life and the universe. I could never persuade numbers to add up consistently in basic sums, but Maxwell could use them to take him all the way to Saturn and *to explain light*. The Cavendish Museum has Maxwell's own desk on display – it is leather-topped, made of a rich red wood, and I touch it reverentially as I leave.

We walk through the modern, high-tech campus, past signs to NanoPhotonics, Optoelectronics and Semiconductor Physics. Could there be another Maxwell exploring light today, I wonder. Are there further breakthroughs of that scale still to come?

What questions are being asked about light here today? I ask my guide. He tells me that the discovery of lasers changed the whole game and made us look at light differently. There are huge discoveries still to be made, he says, but it's all deep in nanoscience and quantum mechanics, and to tell me more he would have to use words like "decoherence" and "wave particle duality".

On my homeward journey, I ponder Maxwell's genius and his ability to demonstrate the nature of light: that light *is* electromagnetic radiation. All well and good, but now I have to ask – there is no way round it – what exactly *is* electromagnetic radiation?

I try the internet and books, but I struggle with the language in which the explanations are phrased. I need a real person, who gets the physics and can communicate it clearly. I ask around – does anyone know a friendly physicist? – and I'm soon introduced to Chris, who is an astrophysicist, which helps. This stuff of stars is where – in the public perception – physics gets sexy, and so astrophysicists are often called upon to communicate beyond their specialist field. Chris is more used to talking about light zooming around the universe from distant galaxies, but he's happy to discuss light bulbs with me. I suggest we meet up in the pub: I'll get the drinks and he'll answer my questions as straightforwardly as possible. It's a challenge, we agree, because the language of physics is more mathematics than words – physicists can speak to each other in equations. I'm better with metaphors, I tell him, and maybe some stories . . .

Physics and a pint, session 1: Electromagnetic radiation

This is just my kind of pub. There's a tangible comfiness to it: the way the evening sunlight falls on the warm wood of the tables; cosy corners; soft halogen lights. I get us both a pint then reach for the paper and pen.

Let's begin at the beginning. What is *light?*

"Light is a wave – an electromagnetic wave. It travels in a vacuum through space. Think of it as a changing electric field, which induces a changing magnetic field, which induces a changing electric field. They generate each other and that forms a wave.

"A wave is a disturbance that is travelling. The easiest form to understand is a wave in air or water. For example: we speak, which disturbs the atmosphere, which disturbs the atmosphere next to it, then next to that, and so on until the sound wave reaches your ears. Because it is a wave, it has a wavelength (the distance between the waves) and a frequency (the number of waves created per second). As these waves travel at the same speed – the speed of light – waves following each other fast are close together, and the waves that are generated slower are further apart: a long wavelength has a low frequency, and a short one a high frequency."

Chris draws a long, wavy line, the waves far apart then – from left to right – getting closer and closer together.

"These waves form an infinite spectrum of electromagnetic radiation. What we call light is the small section of this that is visible to the human eye. These waves form the rainbow of colours from reds to blues. Beyond what we can see, in both directions, are wavelengths that we cannot see but can use. Continue beyond red as the wavelengths get longer and become infrared, and then microwaves and then on into ever-longer radio waves."

How long do they get?

"We don't know. At the low end of the 'usable' spectrum is the exceptionally long wave radio that is used by the US military because long wavelengths can penetrate deeper into the water. The antennae needed for this crosses several states of the US! Beyond this there is no usable phenomena – a low frequency of, say, one wave per second would be bigger than the Earth! There's no way we can interact with that."

And up the way?

"Heading beyond visible light in the other direction, with ever-shorter wavelengths and higher frequencies, we're into ultraviolet light and then on to X-rays and gamma rays. Again, we're still exploring this spectrum and expanding on what we can actually use. Light is increasingly being used in the medical world, with very intense light utilised to penetrate further into the body. Certain wavelengths are not absorbed by cancers, for example, and so cancer can be detected by the way light changes. Gamma rays are used in some medical research to scan inside the body."

Do they go on forever?

"Mathematically speaking, yes, to infinity, but if two gamma rays meet they destroy each other."

I see. Sort of. I'm beginning to get glimpses.

Physics and a pint, session 2:
Measuring and comparing light

What's troubling me, I explain as I set our drinks down on the table, *is not having a language to articulate my experience with light. And whether the comparisons we make between types of light make sense; are the words we use to describe and compare light equally meaningful to different types of light? Light bulbs today are labelled in a series of equivalents: lumens, watts, colour temperature. And I keep coming across "K" or "Kelvin". What does it all mean?*

"Okay, let's start with Kelvin, because it's a straightforward one", says Chris, "and it leads on to understanding incandescent light. Kelvin is just another measure of temperature, like Fahrenheit or Celsius. For physicists, it is the preferred term because 0K is the lowest possible temperature, also known as absolute zero. Kelvin uses a similar scale to Celsius, but with a different zero: 0°C is roughly 273K, boiling water at 100°C is 373K. It's named after another nineteenth-century Scottish physicist, Lord Kelvin. Kelvin is used as a scale of colour temperature."

That's the bit I struggle with – how can colour have a temperature?

"Think of a fire – beginning with a red glow, getting more yellow until it's 'white hot'. That's the gist of it. In physics, to put specific measurements to this, we use a concept known as 'black body radiation': essentially the colour of something hot, which isn't reflecting light from anything else. So all the colour comes from its own emission of light – changing from that glowing red, through orange to white as it gets hotter. Giving numbers to these colours is a way of describing heat.

"There are some fundamentally different ways of making light. The most obvious one starts with the sun, or, more generally, light from hot things – or 'bodies' as we call them in physics. Imagine a metal wire attached to an electricity supply. As more electricity flows through the wire it gets increasingly hot, and the colour of the glow is directly related to the temperature of the wire. If the wire is around 800K or 525°C, it'll show a very faint, deep red. If the wire comes up to 1000K, it becomes dull red, and when it's 1200K it's bright cherry red. In real life, the wire will have snapped by now, but we can carry on because we're in a thought experiment. We crank up the current and at 1500K we see clear orange and at 3000K it's white.

From there the white gets increasingly brighter and eventually more blue. The sun has a surface temperature of a mere 5800K – 'mere' given that the temperature of its core is estimated at 15 million K! But it is only the surface that radiates out light to us."

So this is all about incandescent light?

"Yes. Incandescent light is light originating from heat and these are the colour temperatures, which describe incandescent light, whether from the sun or a fire or an incandescent light bulb. In the nineteenth century, several inventors knew what had to be done to create a lamp in this way: a wire had to be heated by electricity in a protective or near vacuum environment. Thomas Edison was looking at carbon filament in a vacuum, and in his patent of 1879, Edison described the process of making a carbon filament and how to house it inside the glass.

"A continuing story of patents and claims of ownership followed; many improvements were proposed, and some were implemented. The most significant change was when the carbon filament was replaced by tungsten, a material with an exceptionally high melting point of 3700K, meaning it can be heated to a level where it produces a lot of light. Filling the bulb with inert gas or nitrogen extended the life span of the tungsten filament and increased the brightness as a higher temperature was possible."

And a halogen bulb is also incandescent? I've often wondered what the difference is.

"Halogen lamps are a further development of this principle and have been on the market since General Electric's patent of 1959. Halogens were invented to solve a problem with incandescent bulbs: atoms from the filament evaporate as it heats and are deposited on the inside of the glass bulb. Over time, the filament gets thinner and the glass gets less transparent. Trying to resolve this, inventors experimented with halogens (a group of chemically related elements, including fluorine, chlorine and iodine). They found that if you put a tiny amount of a halogen inside the bulb, the tungsten doesn't deposit on the glass but has a temporary chemical bond with the halogen, after which the tungsten is placed back on the filament. In effect, it's a recycling process: tungsten atoms leave the wire, have a fling with halogen and return. Because this process allows higher temperatures, halogen bulbs make a brighter light."

Okay, I'm following so far. The first way of making light is through heat: heat something up and it gives off light, changing colour as it gets hotter, which is how we can use the number of a temperature on the Kelvin scale to describe a colour. This is how incandescent and halogen lamps work. Got it. But hang on a minute . . . The other types of light bulb don't actually get hot, especially LEDs, because they use very different physics to create light.

"Correct."

Yet the whole discussion about LED street lights is phrased in Kelvin. So when we describe the colour temperature of an LED light in Kelvin, are we talking about the equivalent colour that an incandescent light – or a black body radiator – would be at a certain temperature?

"Yes. It is called the 'correlated colour temperature' or

CCT: a description of what our perception of an equivalent colour would be."

And does that work?

"Yes and no. Kelvin, or colour temperature, is just one way of measuring or describing a certain aspect of light, attributing a single number to the source of light and so making comparison with another source easy. A more scientific way of describing the colour of light is by what is known as 'spectral distribution'. Here, we show how the power of light is distributed over the spectrum, showing which colours or wavelengths are represented more strongly than others.

"Going back to the sun – in the white light it produces, all wavelengths of visible light are present. The colour temperature of sunlight before it enters our atmosphere is 5800K and we can draw a smooth curve from reds to violets; this is called a spectral distribution curve."

Chris draws three smooth, curvy lines. "The sunlight gets filtered in the atmosphere: midday sunlight contains more blue and so the curve slopes gently down to red; the evening sunlight slopes up to a big curve of red. As a result, the CCT of midday sunlight is different from the CCT of evening sunlight, and both are different from the actual 'black body temperature' of the sun."

But Maxwell showed how you can make light, the appearance of white light, by mixing just three colours: red, green and blue . . . You don't need the colours in between for the eye to perceive it as white light.

"Indeed. This is where we step from physics to biology.

Because of the way the eye processes light and colour, we don't need all the colours to be present to give the appearance of white light. The eye has different types of light-sensitive cells, including the rods and the cones. The rods are the most light-sensitive, but cannot distinguish between colours. They help us see in low-level lighting environments, where the more sophisticated cones stop working.

"The cones come in three different types. All three are sensitive to light levels, but they differ in sensitivity to different sections of the visible spectrum. Usually they are referred to as red, green and blue cones. The red cones are sensitive to the light with the longest wavelength, the green with the intermediate wavelength and the blue with the shortest wavelength.

"Our eyes and brains collaborate to distinguish a colour by comparing the signals produced by the three types of cones. An orange light will excite the red cones a lot, the green cones slightly less, and the blue ones not at all. Our brain gets these signals – plenty red, some green and no blue – and concludes: orange.

"But we can trick the brain into thinking it sees orange light, by having a red and a green light close together and have them switched on with the red a bit brighter than the green. The signals the eye sends to the brain – again plenty red, some green and no blue – are the same as if the eye was subjected to real orange light. Television and computer screens with red, green and blue pixels use this optical phenomenon to trick us into seeing all colours of the rainbow.

"So, by mixing red, blue and green light and tweaking the proportion of intensity, with respect to each other, we can make the appearance of colours of light equivalent to the CCT: warm yellow, white, orange, cold white, blue, etc. Two light bulbs with the same CCT may have very different spectral distribution curves."

Chris draws the pattern of light from an incandescent bulb, which resembles that of evening sunlight. Then a CFL: instead of the smooth curve, there is a high jagged spike of blue, then another of green, then another of orange. And a white LED appears as a spike then a hump, which makes me think of a rhino: a high, narrow pinnacle of blue and then another more rounded hump beside it rising from green and sloping down to red. I'm intrigued – it's the first time I've seen a visual representation that explains a real difference in the light from different bulbs. So, the relative spectral distributions is one way in which the light emitted from a CFL bulb is indeed a different *stuff* from that of an incandescent or an LED bulb.

And could that be something to do with the problems that some people are having with these lights?

"I don't know. I'm a physicist, not a physician. But any issue of light as we perceive it cannot be separated from the complex physiological mechanisms of perception, in the eye and the brain and the relationship between them."

Where does a "lumen" come into this? In shops that sell light bulbs I've seen signs saying "Think lumen not watts". How does one "think lumen"?

"A lumen is a way of measuring the quantity or the

intensity of visible light emitted from a source. The interesting bit here is the phrase 'visible light': not all the light emitted is visible, and not all visible light is perceived the same way by individuals. Lumen mixes the terminology of physics and physiology to some extent. But it's mostly based on a model of human vision, so is used more in lighting design. It has very little to do with physics. It's a measure of the amount of light that humans perceive, but it treats wavelengths differently: it uses our understanding of human sensitivity to different wavelengths – or colours, if you like – to weight the power of a lumen. So, for example, a red LED of one lumen is emitting a lot more light power than a green LED, because a human eye is much more sensitive to green light."

But is that reasonable? I've come to understand that there are a lot of subtleties in human vision, between individuals and especially in different age groups.

"It's an approximation. All these measurements are approximations and therefore the question arises: when we describe light in a certain way, what approximations do we make unwittingly? What information is missing?"

So what is the best way overall of describing light?

"There is no overall way of describing – or comparing – light. All these terms are useful for certain things, but in other ways they are crude and oversimplified. It has to be oversimplified to be workable for some purposes, but the nuance gets lost and that can be important. The 'feel' of light depends on a number of variables and no single number or measurement can describe them all."

Incandescent

As I understand it, one of the reasons for banning incandescent light is a comparison in "lumens per watt" – how meaningful is that?

"That's open to question: at what point can you reduce something so complex to one number, and is that the thing which is most relevant? Is that what you're actually trying to measure?"

Physics and a pint, session 3:
Going quantum

I'm straight into it this time: *I know that in other types of bulbs – fluorescent, LED, etc – a different kind of physics is used to make light. But what is it?*

"Okay, let's just say that in incandescence, heat shoogles the atoms a bit inside a material and that makes the electrons jump, and it's the process of electrons jumping – and sending packets of energy back and forwards – that makes light. Without heat, you have to find another way of making the electrons jump. Different technologies have found different ways of doing this."

Chris is looking at me intently. "Now, we have a choice," he says seriously. "We could stop here, leave it at that? Or, we could" – there's a dramatic pause – "go on. If we go deeper in this direction, there's no way round it: we're going quantum."

Not long ago I would have run a mile. Quantum. The very word is practically a euphemism for "you are not going to understand this". But I'm getting curious. *Try me . . .*

"This is part of the story, because the quest to understand what light actually is led scientists to go deeper and deeper into an atom. And through it – a portal, if you like – into the world of quantum mechanics. So the first thing to understand is that in quantum theory we're dealing with small. Very, very small. The smallest components that the universe is made of. Okay so far?"

Sure, I've seen small. It's amazing. I used to work at the Botanic Gardens and saw the Scanning Electron Microscope, which botanists would use to study plants at 30,000x magnification. It's like seeing another world: familiar things are transformed into fantastical forms; microscopic pollen grains become spikey footballs or giant sea urchins.

Chris is smiling kindly, patiently. "Yes, that's small, but we're talking a million times smaller. No," – he corrects himself – "maybe a billion times smaller than that.

"It is, in a way, like another world. Imagine we go through the atom portal into a parallel universe, which runs on different lines, where the rules we know don't really apply any more. This world is composed of counterintuitive, infinitesimally small exactitudes with no shades of grey or in-betweens. We're even in a land beyond metaphors – the concepts we use as tools to explain things just disintegrate as soon as they enter this world's atmosphere. And yet, the contents of this universe is what our own and everything in it is comprised of, and so these different rules ultimately shape and govern our own . . . Are you still with me?"

I don't answer straight away, but I'm smiling. For this is

sounding far from the physics I barely knew, and delightfully like some intriguing science fiction. Or, better still, like
a novel by one of my favourite writers, Haruki Murakami,
with a dreamy interchange of parallel worlds that are separate and yet the same, interconnected and influencing each
other in mysterious ways. *Go on, tell me more.*

"So, we describe light as both a wave and a particle at
the same time. To get an impression of this, imagine light
reaching us as small packets of waves. On each packet
travels its captain, Captain Photon. Depending on the
wavelength, Captain Photon carries with him a very specific
amount of energy: a quantum. Nothing more, nothing less:
the shorter the length of the wave, the greater the amount
of energy he carries. A packet of ultraviolet light has more
energy than a packet of blue light, which in turn has more
energy than a packet of orange light."

You can have packets of crisps, I grumble, *but not
packets of waves! I just can't imagine that.* I'm big into
metaphors and they're precious to me. I don't mind people
mixing them, or milking them, but Chris is messing with
them and with my mind.

Chris defends himself: "Like I said, this is a place where
metaphors disintegrate."

*Yes, but this is getting crazy, and I can't see how we can
ever get back from here to light bulbs . . .*

"Call it a ripple, if you like," he says, "or Captain Photon
riding the wave? But the wave is contained in some way,
hence the packet."

But you cannot contain a wave.

"Exactly! That's the duality." As if that explains everything. "Okay, imagine an atom as a massive roundabout with a lot of lanes. The electrons on the roundabout are zooming forever round and around their lanes, and only one, or a few, of the outer ones will ever leave their lane. They change lanes by *jumping* into the next – it's a very precise move without ever straddling two lanes. The outer electrons can be persuaded to move further out if there is space and when they get a very specific amount of energy. They can get this energy by impact with a Captain Photon. When Captain Photon collides with an electron, he is dead and gone. He hands over his energy and ceases to exist. Pow. That light wave is gone. The energy from Captain Photon has to be exactly the right amount to make the jump and move up a lane.

"Now, there is another way for an electron to change lanes. This happens when a wild electron from outside the roundabout comes flying in and knocks an orbiting electron into a lane further out. An electron zooming around in the outer lane is not in a happy place and it will soon return to its former lane. When doing so, it releases the energy that it used to get into the outer lane. The energy is released as light with a new Captain Photon."

Are we ever going to find a link from all this to light bulbs?

"Yes, I'm just coming to that! The energy it takes an electron to move up a lane, and the energy it releases when it moves back, is a property specific to the material. So, for example, let's look at a sodium lamp. Remember the old

street lights, which made everything look the same oran-
gey-yellow colour?"

*I remember them well. If you walked in the dark towards
a village with those lights, you'd see the yellowy-orange glow
above the houses. It always felt welcoming.*

"Well, sodium is a metal where the outer electron can
hop into two outer lanes, which are close together. When
the electron jumps back, it produces a yellow light."

*And how do you knock the electrons into the outer lane
in a sodium lamp?*

"A low-pressure sodium lamp is a glass container (the
light bulb) with neon gas that is so hot that its electrons are
off the roundabout completely: they're shooting around in
the tube. Also, because of the heat, the sodium is vaporised and its atoms are floating freely in this space as a gas.
The shooting electrons originating from the neon push the
electrons of the sodium on to outer lanes. On their return
to the original lane, they emit that characteristic yellow
sodium light. This yellow light is of one two very specific
wavelengths."

*And so everything under it looks the same shade. There
are no colours – even a blue car is just dark yellow. But we
rarely see these street lights anymore. We've got horrible
fluorescent street lights on our road now . . .*

"This phenomenon is also used in fluorescent lights,
but using different materials. In fluorescent lights, instead
of using sodium, the vaporised material is mercury. Inside
the tube, the mercury is vaporised by the heat and when
the shooting electrons hit the mercury electrons, those

mercury electrons move into an outer lane and when they return they emit an ultraviolet (UV or UVR) photon. So inside the tube we are generating vast amounts of high-energy photons. In fluorescent lights, there's an extra step. The tube is coated on the inside with carefully selected materials that have electrons in the outer lanes, which will give off a mixture of more useful colours than UV when they are bombarded by the high-energy photons."

And what about LEDs?

"An LED – that stands for light-emitting diode; a diode is an electronic device used in electric circuitry – uses electroluminescence to make light. Electroluminescence is the phenomenon of certain materials emitting light in response to an electric current. An LED makes light of a very narrow wavelength. This can be done in different colours, but it is most efficient to produce blue light, then the blue can be converted into other colour lights using a phosphorescent material. Similar to the process in fluorescent lamps, high-energy photons (blue) are converted into lower energy photons (green, yellow, orange, red) and, provided the phosphorescent materials are added in the right proportions, different shades of what we perceive as white light can be produced. In LED, the mechanism of generating blue photons is different but they have the final step in common with fluorescent tubes: photons are bombarding a layer of carefully selected materials to create all sorts of colours."

Phewee, I need a drink, and I need to digest all this. I can see how different materials are used to create light in different ways. I'm still struggling a bit with those packets of waves – the idea of light being a particle and a wave.

"Okay . . ." Chris takes a deep breath. "Light isn't *actually* a wave. Nor a particle."

What? I feel like I've been slapped. *That was the bit I had grasped! We'd begun this whole thing with a nice clean sentence: "Light is a wave – an electromagnetic wave."* Chris had said so. Maxwell had said so. Even Wikipedia had said so when I'd been Googling late one sleepless night.

"I guess it's a metaphor," Chris explains. "Perhaps a simile would have been more honest: light is *like* a wave. Or, light behaves in certain ways that are analogous to our understanding of the behaviour of waves. Other times it behaves more like a particle."

There's a silence.

"Pint?" Chris offers.

I'm supposed to be getting the pints.

"I'll get this one."

I think I need a whisky.

The whisky soothes and uplifts. It's a single malt, rich and smooth. I stare into it, tipping it slowly against the side of my glass. This is as close as a drink can get to the warmth of an open fire, the golden tones of firelight, of sunshine. Maybe we should grade whisky by CCT or on the Kelvin scale, or by spectral distribution curves.

"One of the key features of quantum mechanics is unknowability," says Chris in mitigation. "There are certain things we cannot know, not even in principle. The mystery is part of the puzzle. It inherently leads to uncertainty, to things that cannot be measured or explained. Ever. We can understand it well enough to work with it, but not well enough to truly know what the hell is going on. There is no way to figure it intuitively; we have to accept it just is."

Let the mystery be? Maybe it's the effect of the whisky, but I'm smiling again now. If not *terra firma*, this is at least familiar territory, for it's beginning to sound surprisingly like the way Quakers talk about God.

"This is difficult for physicists," Chris says. "We seek to understand, to explain how things work and to answer questions, we want to be exact."

I'm okay with mystery, I say, *I find it very reassuring.*

4

Body and Mind

So there is light: a complex, multifaceted mystery of electromagnetic waves shooting across the universe and across our sitting rooms, encountering matter, being reflected, refracted, absorbed. This stuff of stars and rainbows, computer screens and laser beams, of different intensities of energy.

And there is the human body and mind: a complex, multifaceted miracle of heart beating, muscles pumping, blood rushing, neurons firing, cells dividing – all choreographed in precise coordination within a thinking, dreaming whole. Humans evolved over millions of years – like all of life on Earth – in a specific relationship with light.

What happens when the photons of light encounter the living, pulsing matter that constitutes a person? And now, as light is changing, in both quantity and quality, how does it affect the human mind and body?

Everyone I've met on this strange light journey has been going through the same process of carrying questions around in the hope of finding answers. But we find ourselves heading up blind alleys or round and around in loops that lead us back to where we started. We try the doctor first, then specialists, then organisations representing various conditions, then political representatives – surely *they* must have enough information about the effect of artificial lights to ban one form and enforce another?

But the response is a set range of non-answers: we don't yet know; this technology is new; we're not aware of any problem; there is no evidence; there is insufficient data; this is anecdotal. And so we're left dangling, with our stories floating around in cyberspace, and each of us clutching our own piece of the jigsaw. When we meet one another, these pieces don't fit together. There are too many in-between bits missing, but they give us a glimpse – enough to convince us that we are all part of a much bigger picture.

When I dig a bit deeper, just a bit of online browsing, I find plenty of medical experts expressing serious concern about the effects of artificial light on humans, and this disquiet dates back to long before the incandescent ban. Ophthalmologists, dermatologists and neurologists all publish academic papers in journals and speak at specialist conferences. But how does any of this reach the public or the policy makers? Or even the GPs? And is anyone joining the dots between it all?

I'm frothing with questions for these experts. How can my husband sit beside me at a friend's house and not be affected by the same light as me – why can't he feel it? Why is my light-sensitive friend not affected by CFLs, yet reduced to vomiting and fainting within seconds of exposure to an equivalent bulb that's an LED?

Among the people I'm in touch with who have trouble

with new lights, there are such diverse symptoms described, but they fall into rough categories: skin – rashes and burning; eyes – pain and vision problems; brain – headaches, migraine, confused thinking and anxiety. And it seems reasonable that any of those symptoms that affect the eyes and brain could have a knock-on effect and trigger the nausea that myself and others frequently experience. I decide to keep on digging and see what I can find out, to follow my own questions and see where they lead me.

How light affects skin

I start with skin: the largest organ in the human body, and where the relationship with light is most understood. You can feel it, and it's visible – one day in the sun and you can *see* your skin interacting with light, being changed by it. Concerns about skin health have brought talk of wavelengths and light frequencies into the popular consciousness, after years of public health campaigns about the risks of excess UV radiation from sunlight and especially from sunbeds. Skin was also the most immediate – visible – indication that something was wrong with CFLs. When these bulbs first came on the market, some people reported skin burning, blistering, rashes and itching, even "weeping, bleeding pustules" after close exposure to CFL bulbs.

- Are experts worried about the possible effects of CFLs on skin?

Emeritus Professor John Hawk, of King's College Hospital,

London, is the President of the World Congress on Cancers of the Skin. He has long expressed concerns about close and constant exposure to CFL bulbs. Back in 2008, as dermatology spokesman for the British Skin Foundation, he warned BBC Radio 4's *Today* programme that "fluorescent lights seem to have some sort of ionising characteristic where they affect the air around them. This does affect a certain number of people, probably tens of thousands of people in Britain, who are flared up just by being close to them. Certain forms of eczema – some of which are very common – do flare up badly anywhere near fluorescent lights, so these people have to just be around incandescent lights."

Professor Hawk was an observer at a meeting about CFLs in Brussels in 2011: "The overall feeling of the meeting was that the lamps had a number of potentially adverse effects, mostly for abnormally photosensitive subjects but also somewhat for normal ones, on both skin and eye."

In 2012, he summarised the effects of CFLs on skin in a statement: "They are safe for normal people at a distance (say, at least two or three metres, to be sure) but close up, they can cause sunburning in some cases, and constant exposure can add up with sunlight (and sunbeds if used) to increase the risk of skin ageing and cancer. In people with certain skin and generalised diseases, however, they can also cause very unpleasant effects, either a flare of the disorder, or a very unpleasant skin sensation.

"These flares in lupus [an autoimmune disease] and seborrhoeic eczema may sometimes just be extremely unpleasant sensations with no visible skin change, and this may on occasion lead non-dermatologists to think patients are making the story up, but I have seen many such subjects who do not know each other with exactly the same stories, indicating the conditions are real."

• What is it about CFLs that is potentially harmful to skin?

CFL bulbs emit ultraviolet radiation, and one of the issues is that individual bulbs vary widely in the intensity and the type of UV that they emit (ranging from UVA, closest to the visible spectrum, to UVC, the furthest away). But there is no labelling to inform consumers what they are being exposed to. There are aspects about CFLs that make this more of a problem than in ceiling-mounted fluorescent strip lights. For a start, there is the physical proximity to skin when CFLs are used as reading lamps and on office desks. The lack of heat means people get closer to them than they would to an incandescent bulb, and the poor light they emit often draws people closer still.

A further issue is the design: the way CFLs are curled and looped to fit a narrow tube, so that they resemble a traditional light bulb shape, places stress on the phosphor coating, making cracks far more likely than in the larger, ceiling-mounted fluorescent tubes. These cracks "leak" UV at higher intensities. One study, which closely examined a wide range of bulbs, found cracks in all the phosphor coatings, thus exposing anyone sitting close by for a length of time to unsafe levels of UVR, including the more harmful type: UVC. "Our results confirm that UV radiation emanating from CFL bulbs (randomly selected from different suppliers) as a result of defects or damage in the phosphorus coating is potentially harmful to human skin."

The British Association of Dermatologists issued a statement on CFLs, which explains the different types of UV emissions from various lamps: "Although the traditional tungsten [incandescent] lamp emits measureable amounts of UVA, the UVB levels are extremely low and the heat from this lamp limits exposure at close distance. In contrast, the CFL emissions are widely variable, depending on the individual lamp. The single envelope CFL [a coiled bulb with no additional outer casing] in particular has been shown to emit significant quantities of UVB in some cases and also measureable amounts of UVC. Light-emitting diodes (LEDs) have virtually no UV emission."

• Is it only those suffering from specific light-sensitive conditions that are affected?

No, because skin itself is sensitive to light, and – as we're so often told in guidance about sunbeds – excess UV can damage anyone's skin, with the very young and old being particularly susceptible.

The dermatology department at Dundee University Hospital has been studying the effects of CFLs and other new lighting technologies. A 2013 paper from this research stated that "UVR from CFLs can aggravate the skin of photosensitive and healthy individuals when situated in close proximity. Double envelope lamps reduce this risk."

Double envelope lamps are when the coils are contained within an outer casing so that they look more like incandescent bulbs.

Official UK government advice is that we avoid spending more than one hour a day closer than 30cm to a single envelope CFL.

• So are LEDs better for skin?

It seems so. The Dundee study concluded: "LEDs offer a safer alternative light source that eliminates the risk of UVR-induced skin erythema [reddening of the skin]."

But it begs the question "safer than what?" Because the succession of light bulbs went from incandescent to CFL to LED, we tend to compare LED with its immediate predecessor of CFL, rather than with incandescent lighting.

There is also the question of better for whom? The website of the charity Lupus UK says that LEDs "emit no UV, which is a good point, BUT some very sensitive people cannot tolerate any of them."

• How we can accurately assess the long-term effects of any new technology?

I didn't find any answers to this one.

• What's the experience of people with pre-existing skin conditions?

Jess has lupus (*Lupus erythematosus*), an autoimmune disease that occurs when the body's immune system is overactive and attacks the tissues and internal organs.

It can affect many different body systems, including the brain, heart, lungs, blood, kidneys and skin. According to Lupus UK, light sensitivity is a common symptom, affecting about sixty per cent of patients, and it varies greatly in severity and in the type of sensitivity.

A former civil servant from Leicestershire, Jess had to leave her job as the lighting in her workplace made her very ill. "I am unable to go near CFLs for more than a few seconds," she says. "The only way to describe the sensation is feeling 'fizzy inside'. I get intense itching all over, like my blood has got ants in it, far deeper than a skin itch. My pulse starts to race, my skin goes red and feels like it's burning, and I struggle to breathe. Then I feel horribly fatigued – it sucks all the life out of me. I can't sleep or lie still due to the itching, then it triggers the usual lupus symptoms: joint pains, headaches, and my knees feel like they're on fire."

Jess found her symptoms worsened during the years it took to get diagnosed. She has seen four dermatologists who didn't acknowledge a problem with lighting. "I'm now seeing a professor at a photobiology clinic – the first medic who understood what was happening and made me feel normal. She said the exposure to artificial lighting was dangerous as it could trigger a lupus flare that may attack my internal organs. So I was unable to continue in employment."

UV light is known to trigger symptoms in lupus sufferers, but Jess explains that the disease doesn't necessarily recognise the boundary between blue light and UV with the same precision as science does. The tolerance to different wavelengths of light varies between individual sufferers – Jess finds hers is even different to her sister, who also suffers from lupus and photosensitivity.

"I still have problems with LED, which is not supposed to emit UV, but the blue light is still a problem. It took five years to source an anti-UV screen for my computer because there is a lack of evidence that they emit UV light and therefore a lack of products to protect light-sensitive people, but that exactness is not how the condition responds.

"When people say, 'You'll be fine with LED', it makes me want to scream. Someone who knows nothing about how life is for lupus sufferers has decided it's okay. Now cars have changed, street lighting has changed, it's far more debilitating – it's a daily nightmare just trying to get out and about."

• What other skin conditions might new forms of light exacerbate?

When the EU banned incandescent lighting, charities representing people with light-sensitive skin conditions lobbied for an exemption, saying that the new lights aggravated or inflamed their conditions, sometimes causing severe discomfort and even permanent damage. As well as lupus there is xeroderma pigmentosum (XP), a very rare disorder in which the genetic material in the skin is unable to repair itself after damage by UVR, and so exposure can cause high risks of skin cancer. Sunlight is dangerous to XP sufferers and they must cover their skin to avoid direct contact. In France, sufferers are known as *enfants de la Lune*, and support groups in France and the UK organise night-time camps and events for children affected. A support group in the UK campaigned for a continued supply of the low-power incandescent lighting (40 watt) that is essential for domestic life.

There is a severe form of seborrhoeic eczema where the skin becomes intolerant of light, including sunlight. In her 2015 memoir *Girl in the Dark*, Anna Lyndsey describes her experience of this condition where any exposure to light, even through clothing, results in an unbearable burning sensation "like a blowtorch held against my skin". She had to leave a busy work and social life in London and live for months, sometimes years, in a dark room, only able to venture outside at night under the orange glow of sodium street lights. When the local council revealed its plans to replace these with new bright fluorescent streetlights, she realised: "If they install white lights, I will never be able to leave the house. If they install white lights behind my back fence, I will never be able to use my garden."

How light affects eyes

I turn my attention to another organ that is exquisitely sensitive to light: the human eye. I treasure my eyesight. While most organs tick along, unnoticed and unappreciated unless something goes wrong, I've always been aware of my sharp sight and value it greatly. Never more so than when whale watching; I get a kick from being the first to spot a distant splash and being able to differentiate a faraway dorsal fin from the shadow of a wave. So of all the stories I hear about problems with new lighting, those concerning eyes worry me the most. The descriptions of pain make me wince, and discussions about permanent damage to eyesight terrify me.

As artificial light seems to become ever brighter and more intense, three issues keep arising: the alarming and immediate pain this is causing some people; the effect on vision and some people's ability to see; and the possibility of real long-term damage to the fabric of the eye.

The main issue here seems to be with blue-rich LED lighting. With the spread of LED street lights, car head-lights, road signs and shops signs, the conversation has moved from a sensitive few to a more mainstream concern. This is partly aesthetic, as lights that used to twinkle now seem to glare, but it's also an issue of discomfort and safety.

• Is blue-rich light a problem for everyone's eyes?

Borek Puza is a statistics lecturer from Canberra, Australia, who writes of his concern about the dramatic increase in light intensity that has occurred worldwide in the last few years, saying he feels "appalled and violated" by the ubiquity of "brutally bright light": "Eye-gougingly bright lights, both LEDs and CFLs, are popping up and fast becoming the norm virtually everywhere: on bicycles and helmets, inside homes, offices, shops and malls, on walls, roofs, lamp posts and bollards, at sporting fields . . ."

Borek wonders how others can barely notice an

intensity of light that he finds unbearable. As with skin, the question arises again – is this an issue where only people with some previous problem or particular sensitivity are affected? Or are these people the "canaries in the coalmine" heralding wider implications for everyone? Borek suspects that the current assault of over-bright light may be causing widespread harm to eyes, slowly damaging retinas, but that most people won't realise this until more of the population is affected, by which time it will be too late: the damage done.

"It is just possible that the collective insanity which is lightmare, both on and off the road, will continue to escalate until an epidemic of irreparable eye damage, creeping up on people in subtle stages without them noticing, has already swept the world. When the penny finally drops, and this is finally acknowledged as fact by medical and government authorities, following numerous scientific studies, it will be too late. If this theory is true, lightmare is destroying the world in at least two ways: by degrading the world as we see it, making it ugly and painful to look at; and by destroying the very eyes that see the world."

It is, he says, just a hypothesis and he would be happy to be proved wrong.

• Can blue-rich lighting actually damage the eye?

In 2010, the French government released a report on health issues to be considered with respect to LED lights, saying that "these new lighting systems can produce intensities of light up to 1,000 times higher than traditional lighting systems, thus creating a risk of glare. The strongly directed light they produce, as well as the quality of the light emitted, can also cause visual discomfort."

It expressed particular concerns about the effect of blue light in LEDs on the eyes of children: "The principal characteristic of diodes sold for lighting purposes is the high proportion of blue in the white light emitted and their very high luminance ('brightness'). The issues of most concern identified by the Agency concern the eye due to the toxic effect of blue light and the risk of glare. The blue light necessary to obtain white LEDs causes toxic stress to the retina. Children are particularly sensitive to this risk, as their crystalline lens is still developing and is unable to filter the light efficiently."

A 2012 study from Complutense University in Madrid warned that continuous, prolonged exposure to LED lighting could cause irreparable damage to the retina of the human eye. "Eyes are not designed to look directly at light – they are designed to see *with* light," said Dr Celia Sánchez-Ramos, who led the study. Modern humans have their eyes open for roughly 6,000 hours a year, she explained, and during that time they are increasingly exposed to artificial light, on screens as well as indoor and outdoor lighting. The study looked at human retina cells in vitro, and showed that maximum damage was observed in cells exposed to blue LED lighting.

After nine days of exposure to blue or cool white LEDs, the retinas of rats showed evidence of "retinal damage and cell death", according to a 2014 study published in *Environ-*

mental Health Perspectives. The researchers attempted to simulate exposure to domestic lighting, and so rather than shining a light directly at the eyes – as earlier studies had done – they set the light source on a rack above a cage in which the rats could run freely.

You might ask how closely in vitro studies, or experiments on rats, relate to living human eyes. Close enough, say the scientists involved, to know that we need to know more. Far more. Professor Chang-Ho-Yang, an ophthalmologist from National Taiwan University's College of Medicine, explains: "Neuronal cells are incapable of repairing themselves or regenerating after damage. This makes it important to pin down mechanisms of injury and link them with clinical studies matching the conditions under which people will ultimately be using LED lighting."

- Does blue-rich lighting affect our ability to actually see?

Bright LED headlights have become an issue of "an unwanted, new road safety risk", says the RAC, whose survey in 2018 revealed that sixty-five per cent of motorists have been dazzled by LED headlights, even when they are dipped. The survey showed that the effect of dazzle was similar among drivers of all ages, and that there was quite a difference among individuals in the time it takes to recover. While some can see clearly again in less than a second, most take up to five seconds to regain their vision, and one in ten respondents take ten seconds before their vision is back to normal. Think of this in terms of stopping distances: a driver dazzled for five seconds while travelling at 60mph would cover a distance of 134 metres, more than a football pitch, before being able to see properly.

And it's not only a question of motorists being blinded by oncoming vehicles. The problem even extends to the drivers themselves of cars with strong LED lights. A columnist in the *Goodwood* racing newsletter describes driving a new Range Rover at night: "With headlights bright enough to disturb sleep patterns in the next country, but beyond the intensely illuminated beam pattern I could see hardly a thing . . . My eyes simply couldn't decipher what lay beyond the blaze."

- What do ophthalmologists have to say about the loss of incandescent lighting?

Ophthalmology Professor John Marshall, of University College London, hit the headlines in 2014 when the *Daily Mail* reported that he was stockpiling a lifelong supply of incandescent bulbs. This was, it reported, due to his concerns about the ultraviolet light in CFLs increasing the risk of macular degeneration and cataracts – the headline said, "to protect against blindness". Speaking to *Optician* magazine shortly afterwards, he qualified his comments, saying he wouldn't use "blindness" in that context, and that these wavelengths are not going to cause macular degeneration or cataracts, but they are potential risk factors. He was concerned, he said, about "blue and UV biologically unfriendly sources being used to replace biologically friendly sources".

In a series of opinion pieces in the *International Review*

of Ophthalmic Optics he explains how his interest began during his PhD with the Royal Air Force to investigate the potential damaging effects of lasers on the retina. His work contributed to international codes of practice, used by the United Nations and World Health Organisation, to protect individuals against such acute, intense light. The potential hazard of blue light is treated as a special case in all such work.

Light is light, he says, whether generated in a laser or an incandescent light bulb: all light sources emit photons, and his interest is in the interaction between these photons and biological tissue. This can be a powerful tool for healing; Professor Marshall pioneered laser eye therapy and his inventions are used across the world. He also became increasingly interested in how excessive exposure to light – whether in level, power or duration – had the potential to damage the visual system. His studies showed that the retina was most sensitive to the short wavelengths in blue light, hence his concern with CFLs and LEDs with their high-intensity light. He explains how this can "exacerbate the ageing process in our eyes", as it can with skin. But while the cells of the skin's surface constantly renew themselves, the cells of the retina don't. The photons from the blue region of the light spectrum, from turquoise-blue to violet-blue, have enough energy to change the molecules that absorb them, he explains. Therefore we need to pay attention to the potential hazards of any blue-rich light source.

Professor Marshall describes incandescent bulbs as

"biologically harmonious", in contrast to LEDs, which "have much higher spectral emissions in the blue and at levels that may require attention over cumulative exposures during a human lifetime".

"There should have been much more consultation with the biological vision community before these biologically unfriendly sources were introduced," he says. "In my opinion there should have been a committee of experts assessing the health hazards of low-energy lighting before they became available in the marketplace and certainly before incandescent bulbs were banned!"

How light affects the brain: headaches and migraines

Everyone seems to know someone who gets headaches from the lighting in an office or in a workplace. From what I hear, it seems headaches and migraines are the two most common complaints about exposure to new forms of artificial lighting. Every day on social media I see descriptions of headaches caused by fluorescent strip lights, CFLs and LEDs: "Feels like projectiles are being shot through my eyes and surging up into my skull"; "Feel like my head got turned into a Swiss cheese at a gig last night"; "Like a serious hammering to my brain, like my head is going to explode".

• How does light impact migraine sufferers?

A report on migraine and light sensitivity by the charity Migraine Action explains that light sensitivity is common among migraineurs – both as a trigger for some, and as a symptom of migraine for many more people. Migraine sufferers are particularly sensitive to each end of the light spectrum – both the red and the blue. As well as issues of the electromagnetic frequency of light, common migraine triggers are glare and flicker.

Recent research shows a complex relationship between migraine and light. It's not just a question of intensification of the headache pain that causes migraneurs' aversion to light. Newly described pathways in the brain carry the electric signals of light to the areas of the brain that regulate emotions and the autonomic nervous system (the part of the nervous system responsible for the control of the bodily functions that are not consciously directed, such as breathing, heartbeat and digestion). During a migraine, light triggers more changes in these areas, and can induce negative emotions, such as sadness, irritability and fear.

• What is it about new lights that cause headaches and nausea?

Professor Arnold Wilkins from Essex University is an expert on visual perception. He feels that flicker is "the most likely explanation" for many problems with artificial lighting.

"Flicker disrupts the communication between the eye and the brain," he explains. "This causes neurons in the brain to fire at the same time – flicker at high frequencies interferes with the control of eye movement, producing little patterns every time you move, which makes it difficult for the brain to work out what is going on.

"Headaches, nausea and dizziness can all be caused by flicker. We have long known that low frequency flicker (say, fifty per second) can cause seizures – when televisions consisted of cathode ray tubes, the flicker was the primary cause of photosensitive epilepsy.

"When CFLs first came in, they were used with high frequency electronic circuits to drive them, and so there was a lot of residual flicker. LED can be much worse, but also much better. You can have really good light with LED when controlled, but it is a question of the quality of the electronics, cost and design. Cheap ones flicker more, and the light they give is poor. But when selling lamps to an unsuspecting public, there is nothing to go on for those purchasing a lamp – the information is not supplied as to what extent it flickers. I feel there is an obligation upon people who light public places to do so in ways that are not flickering."

Professor Wilkins' research has shown that visible flicker causes the most obvious health impacts, with immediate effects including headaches and epileptic seizures – which can be triggered by flicker in individuals with no previous history of epilepsy. However, less obvious and less immediate effects occur from flicker that is invisible, including headaches, eye strain and malaise. You don't have to be able to see flicker for it to have a negative effect. Flicker that is too rapid to see can still be perceived by the brain. "Flicker is subtle," Professor Wilkins says. "You don't have to be aware of it consciously for it to be a problem. And some people are more sensitive to flicker than others."

- Are there other aspects of new lights that cause headaches and migraines?

"I've encountered LEDs that are certified flicker-free, but they still cause me to experience disabling symptoms," says Glen, an engineer from Oxford who has struggled for years with the negative effects of sensitivity to LED lighting. "People say, 'Ah, it's the flicker. We can fix that!' Others cite blue light as the cause. Some talk about intensity or dispersion issues. All these concerns are relevant, but I know intuitively and scientifically that it is a lot more complicated than any one issue.

"There is a lot of work going on to 'solve' the problems of LED lighting, whether by removing the flicker, reducing the level of blue light, or smoothing the spectrum. But each of these issues contains layers of complexity. In low-energy lighting, we're dealing with the combination and cumulative effect of many irregularities such as these. We don't truly understand the properties of light yet, let alone its interaction with physiology. It may well be that some problems are to do with aspects of light we haven't even discovered."

How light affects the brain: thinking

As well as headaches, I hear a lot about "brain fog" and troubles with thinking under new lighting. And what *is* that sense of wrongness that's felt by so many but hard to locate in any particular aspect of body or mind? What is it I'm feeling when I realise – through a physical sensation – that there's a CFL in the room? I feel it like a prickling presence of something toxic.

Then there's something else, that can linger for days after an encounter – an existential bleakness that puts me in mind of the "foul, soul-sucking" Dementors in J. K. Rowling's *Harry Potter* books.

• Light is the visible section of the electromagnetic spectrum. But could some of the reported effects on the brain be due to exposure to frequencies beyond the visible spectrum?

After being "overwhelmed with viewer calls and emails telling us these bulbs were making you sick", Canadian TV news programme 16x9 launched an investigation into UVR and other invisible electromagnetic emissions from CFLs. The Canadian government had said electromagnetic fields (EMFs) emitted from CFLs were well within the international safety limit and so "not an issue of health concern", but researchers from the Swiss government were concerned. CFLs were within the limits, they told the Canadians, but at the top end and there was significant variation in the levels, even within bulbs that look the same.

However, these international limits are set by measuring emissions in the air around the device, not how they affect a human body. The distribution of the electric field changes with a person present, and this needs to be taken into account. The film showed energy moving through the human nervous system, down the spinal cord and into the arms and legs, when someone is exposed to a CFL. A Swiss scientist concluded: "It gives the indication that they should actually be tested before marketing."

• Who or what sets these international limits for electromagnetic emissions?

The International Commission on Non-Ionizing Radiation Protection (ICNIRP), based at the German Radiation Protection Agency and recognised by the World Health Organisation (WHO), sets limits for occupational and residential EMF exposure, based on biological effects that have been established to have health consequences. WHO is currently working to "harmonise" EMF standards because they vary greatly around the world.

What I don't understand, and what the Canadian film didn't query or explain, is how legal limits make sense if they are set per product and don't take the cumulative effect into account? If one CFL is near the limit, what about a room containing twenty-seven of them? Or if you're sitting in an office under CFLs, between computers, wireless printer and headsets?

How light affects the body clock and mood

When I began to look into the effects of artificial light on health and well-being, I created a file split into sections – skin, eyes, brain, mind, etc – but the information I found wouldn't always cooperate with my divisions. I soon realised that the interface of light with mind and body was more complex and interconnected. Try getting a file divider between neurology and psychology and you find yourself stuck in a whole basketful of chickens and eggs: neurons and chemicals, thoughts and moods . . . which causes which? And how does light affect it all?

Nowhere is this subject more intertwined than with the eyes and the brain. I used to think the eye was just a "window" from which the mind can look out to "what is there" in the world outside. But no. What we "see" is a construct of our brain's visual processing system, which has developed over millions of years of evolution to focus on the most useful information for our survival. Our visual perception of the world is created by an astonishingly intricate collaboration of these organs, collecting information, responding to colour and form, turning the image upside down, sending it on to the brain to flip back around again, interpret and process and respond to.

The light that enters our eyes contains too much information for our mind to meaningfully consider. It is deciphered by a series of algorithms, illusions and distortions – a process we are still learning about. Moreover, it's one that has profound effects on how we function on a day-to-day basis, and it helps determine everything from our most basic needs – such as when we wake up and go to sleep – to our most complex, such as emotional well-being.

• How does light get from the eye to the brain?

Dr Annette Allan is a neuroscientist at the University of Manchester. I attend a presentation, "From the eye to the brain", in which she explains recent developments in our understanding of this process and the implications of changes in light for our health as a whole.

Humans – like all other life forms – have evolved mechanisms to synchronise behaviour with the changing daylight that surrounds us. The body's main light-sensing organ is the eye and for a long time it was thought that the way eyes detected light was through a system of rods and cones, which are specialised cells at the back of the retina. As we saw in Chapter Three, the rods are responsible for vision in low light, with a focus on shape and form, whilst the cones deal with colour perception and image formation.

Back in the 1920s, it was noticed that blind mice without rods and cones still responded to light – their body clocks were still set by the pattern of light and dark – which suggested there must be some other way of detecting light that was separate from vision. It wasn't until 2001 that this mechanism was understood when other cells in a slightly different part of the retina were identified, and given the snappy title of "intrinsically photosensitive retinal ganglion cells", or ipRGCs. It's an awkward acronym, but hold tight to those first two lower-case letters: "intrinsically photosensitive". This is about light – these cells' very purpose is to receive and respond to light.

These ipRGCs are a crucial piece in the jigsaw of our understanding of light and physiology. They detect how bright the environment is and send this information to the suprachiasmatic nucleus in the brain. In turn, this stimulates or suppresses the hormone melatonin, which has a powerful role in our bodies and minds as it is involved in synchronising our circadian rhythm, or body clock. And the body clock does not just regulate the cycle of sleep and waking, but affects all the physiological rhythms and cycles in the body, including metabolism, bowel movements, body temperature, cell renewal and blood sugar regulation.

• What happens when we alter the quantity or quality of light?

Dr Allan shows an image of a smashed clock – splinters of broken glass, springs and cogs scattered around. It has stayed in my mind ever since. "We're dealing with a broken clock," she says. The spread of artificial light has taken place in a tiny sliver of time in terms of evolutionary history. (And the recent changes in artificial light even more so.) The lighting environment in which we are immersed is utterly different to the steady rhythm of day and night in which we evolved. And because the workings of our bodies are programmed by light, this broken clock can upset everything, resulting in a wide range of health problems, including sleep disorders, weight gain, diabetes, reduced immunity to disease, viral infection, even increased risk of substance abuse.

Much of the technology we use today, in TVs, computer screens and artificial lighting, was developed without this understanding of how our internal systems of light-detection work. Dr Allan's research asks whether we can design lighting and visual displays "with melanopsin in mind". Melanopsin is the photopigment in the specialised eye cells, which send signals to the pineal gland to regulate melatonin levels. So Dr Allan and her team are investigating ways to control this mechanism separately, so that the eye can detect an image without triggering changes in melatonin and the subsequent impact on the physiology as a whole.

It's a challenge, Dr Allan explains when I talk to her after the presentation, to convince manufacturers and legislators. "The understanding of how visual displays work and influence our vision is based on an understanding of vision from the 1930s and that hasn't been updated, for example, by the discovery of ipRGCs. A lot of engineers and manufacturers think that vision is 'solved', and the way lighting and displays are designed does not take into account that our understanding has changed."

And presumably will continue to change?

"It definitely will, without a doubt. In science, we can never assume that our understanding of something isn't going to change. Our visual system is a brilliant example of that because it's a relatively old area of scientific research and people have been working on it for a long time and then we have a realisation of an entirely different way of responding to light. Further understanding will come from the big questions on the power of light on different types of physiology, something we're still only scratching the surface of."

• Just how important is the body clock?

In 2017, the Nobel Prize in Medicine was awarded jointly to three American scientists for their discoveries on the molecular mechanisms that control the circadian rhythm. That we have a body clock had long been understood, but how does the clock keep time? The scientists isolated the gene, which encodes a protein, which accumulates during the night and is degraded during the day, creating "self-sustaining clockwork" inside a cell. Their work was on fruit flies, but applies to other "multi-cellular organisms" adapted to this revolving planet – ourselves included. This clock controls a large proportion of our genes, they said, and the message from the Nobel Laureates is that we mess with it at our peril. It regulates our hormones, blood pressure, body temperature, feeding behaviour, and sleep, and "chronic misalignment" increases our risk of various diseases.

• Is this broken clock just about blue light?

Artificial light sources rich in blue light – including computer screens, CFLs and much LED domestic lighting – excite the ipRGCs at the wrong time of day, stimulating wakefulness and suppressing the hormones that make us sleepy.

While light in general can stimulate the ipRGCs, blue light is the most powerful. A study at Harvard Medical School compared the effect of exposure to green and blue light and found that blue light suppressed melatonin for twice as long and shifted circadian rhythms by twice as much. Sleep suffers. "Worse, research shows that it *may* contribute to the causation of cancer, diabetes, heart disease, and obesity."

• Is there really a link between light and cancer?

The International Agency for Research on Cancer, part of the World Health Organisation, has described shift work that disrupts the circadian rhythm as "probably carcinogenic to humans". Recent studies have shown a significant increase in the risk of breast and prostate cancer, both hormone-related cancers, linked to exposure to blue-rich artificial light at night. The study used images taken from the International Space Station to evaluate the levels of blue-rich light from LED street lighting in our cities and correlated them with incidence of cancer.

Certain drugs used to treat breast cancer are rendered less effective when the patient is exposed to artificial light at night, according to a 2014 study published in the journal *Cancer Research.* The researchers suggest that light at night affects breast cancer in two ways: it inhibits the nocturnal production of melatonin, which itself inhibits breast cancer growth; and it renders the drug Tamoxifen less effective. Darkness during the night could be the key to the success of breast cancer treatment, says the research. Night-time light "may represent a unique and previously unappreciated risk factor that could account for some forms of intrinsic and possibly acquired Tamoxifen resistance and may even lead to a shortened survival time and even a decreased survival rate".

But I also find a 2018 study published in the *British Journal of Cancer*, which analysed data from over 100,000 UK women and their levels of exposure to light at night and found no association between light levels at night and breast cancer risk. "By more accurately assessing the light levels women actually experience in their homes, this study provides the strongest evidence to date that exposure to light at night does not increase breast cancer risk," said the Chief Executive of Breast Cancer Now, which funded the study.

• Isn't blue light supposed to be good for your mood?

A morning blast of bright and blue-rich light is a long-established tool used in dealing with seasonal affective disorder (SAD), now seen as a form of depression. The market is awash with "SAD lights", claiming to "improve mood, energy and focus", "to put you in a better mood and make you feel more awake" and to "deliver 100,000 lux as far away as 35cm . . . to lift mood, restore natural energy in just 30 minutes a day".

A *Which?* report says, "SAD lamps shine very bright, cool light. When this light hits the retina at the back of your eye, it sends nerve signals to parts of your brain, affecting your chemical and hormone levels, and improving mood."

The current NHS guidelines say it is not clear whether light therapy is effective.

• If light alters the chemistry in the brain, can it trigger anxiety?

What is that urge to "get out, get out" that I experience when I encounter a CFL? It feels deep and animal-like, instinctive. Could certain types of light be in some way stimulating the amygdala region in the brain that deals with fear and triggers the body's "fight or flight" response?

I once asked my psychologist if it was possible that light could physically cause anxiety, and she just gave me a look that silenced further discussion or explanation. But now I type "amygdala" and "light frequency" into a search engine and come straight to a US study published in 2010, which says that the spectral quality of light has an acute and direct influence on the way the brain processes emotions, and that "the amygdala, a core component of the emotional brain that receives sparse direct projections from ipRGCs, is one of the brain areas acutely affected by changes in ambient light".

I dig further, and discover the work of US biology and neuroscience professor Samer Hattar. His work reveals the direct role of light, through ipRGCs, on mood and cognitive functions as well as the body's physiological processes. The wrong type of light at the wrong time may be responsible for some cases of depression, raised stress levels and problems with learning and memory deficits – and directly, not just by altering your circadian rhythm and sleep.

"I feel like your grandmother, because this is very simple advice," he says apologetically in a podcast from the Brain Science Institute at Johns Hopkins University in the US. "We think that you should be exposed to bright light and go outside in the day and avoid very bright light or blue-shifted light at night." It's okay to read in bed, he says, but do so under "red-shifted white light" so as not to affect mood and cognitive functions. This means avoiding fluorescent light and screens from phones and laptops and instead using light sources such as candles and incandescent bulbs.

• Could the absence of certain frequencies also be an issue?

While most attention is focused on the potential problems of excess blue and ultraviolet light, questions are now being asked about other electromagnetic frequencies, which may be causing health issues by their omission from artificial light. Thinking back to those spectral graphs which show light as a series of smooth curves in sunlight and incandescent light, and spikes and gaps and humps in CFLs and LEDs – could it be that the gaps are a problem as well as the spikes? Recent findings in cell and molecular biology show that red and infrared light can have important protective effects on tissues and organs, including enhanced repair and regeneration processes in the retina.

•

As I research, it becomes increasingly clear to me that light has such fundamental effects on our bodies and minds. I keep thinking that surely – *surely* – all this and more was researched and tested in rigorous clinical trials before the new forms of artificial lights were put on the market and the alternatives banned, and before this new technology spread at such astonishing speed through our streets and cities and sitting rooms?

There *are* EU committees that have been tasked with assessing the health impact of changes in light. At the time of the incandescent ban this was the Scientific Committee on Emerging and Newly Identified Health Risks (SCENHIR). Its remit was to identify, assess and advise the EU on risks to both human health and the environment. It was asked to determine whether the claims about energy-saving lamps from light-sensitive citizens' associations were valid, and, if so, which aspect of light was responsible and how many people would be affected. Its report in 2008 acknowledged that fluorescent light could be a risk factor for people with certain skin conditions "that are exceptionally sensitive to UV/blue light exposure", and estimated this number as around 250,000 EU citizens.

For the wide range of other health conditions investigated, including migraine, epilepsy, retinal diseases and autism, the report summary draws an overall conclusion that the claims of CFLs aggravating symptoms are "not supported by scientific evidence". Read the detail, however, and it's not such a clear-cut picture. There is a variety of wording on the theme of "we don't yet know": it is "not yet analyzed" if low flicker frequency induces epileptic seizures; a negative influence on autism "cannot be excluded"; problems with fluorescent lamps "are not investigated but are very unlikely" with regard to fibromyalgia; blue light "may be harmful" to those with retinal disease.

SCENHIR produced another report in 2012, on the "Health Effects of Artificial Light", in response to concerns not just about CFLs but other forms of artificial light. It states that some CFLs, if used close by for a long time, emit UV radiation that can exceed the recommended limits set to protect against skin and retinal damage. It also mentions that some light-sensitive individuals report that halogen lamps and LEDs aggravate their symptoms. It identified several knowledge gaps, including manufacturers' data on the light spectrum of each model, health effects of flicker, long-term exposure on the retina, disruption of circadian cycles, and risk of skin cancer.

SCENHIR was succeeded by the Scientific Committee on Health, Environmental and Emerging Risks (SCHEER). In 2018, it produced a report into the potential health effects of LED lighting. It's a weighty document, again written in that curious non-committal language where science meets civil service, the sentences peppered with qualifying clauses: "however . . ."; "although . . ."; "but . . .". This makes it hard to differentiate between reassurance and dire warnings. It all depends on how you read the many instances of "may" and "can be" and "has been suggested that".

"Some LED emission spectra may induce photochemical retinopathy, which is a concern, especially for young children," we're told, and, "Blue radiation directly from bright, cold white light sources . . . may represent a risk for retinal damage." Likewise "older people may experience discomfort", and so on. It recognises that early-to-market LEDs "had a significant blue emission" but reassures us that further research is improving LED lamps to make them similar to incandescents.

I learnt from this report that LED lamps actually deteriorate over time (or rather, "may change"). As the phosphor coating degrades, the white LED may progressively release more short wavelengths – blue light. Therefore, they "might" cause more blue light exposure, which "has been argued" to accelerate ageing in the retina, which in turn, may "possibly play" a role in age-related macular degeneration.

What strikes me most is the sense that we don't yet know the long-term effects, and cannot know them in such a new and rapidly developing technology. Many adverse effects of light on skin and eyes are cumulative, building over decades. The SCHEER review of published research identified significant gaps in our understanding. "There is insufficient knowledge" about actual exposure to people of optical radiation from LED. Brightness, flicker and stroboscopic effects "need to be addressed in further studies". Bright LED headlights in vehicles "should be investigated to determine if there are potential adverse consequences for increased accident rates". Cumulative exposure "should be considered" and "further research should be done" on age-related macular degeneration. Meanwhile, "it is not yet known" about the long-term effects on the circadian system, and so "remains to be investigated".

Meanwhile, we are all the unwitting – and some unwilling – participants in a global experiment on the interaction between light and human minds and bodies. It's an experiment that is unregulated, unsupervised and seems to be rushing ahead of itself, out of control. Our children are among the guinea pigs, their eyes subjected to light that "may" be harmful, "may" cause permanent damage to developing retinas. Instead of waiting for evidence of a problem – which may take years as public pressure tends to build slowly, and because it takes time for scientific studies to be initiated, funded and undertaken – surely we should begin with proof that a product is safe?

What I've learnt from all my digging is that there is a striking amount of individual variation in the responses of human bodies and minds to light. We differ – even within the rods and cones and those ipRGCs in our eyes, in the pigments in our skin, and in the susceptibility of our brains to flicker. When limits are set, or regulations created on any aspect of electromagnetic radiation – whether visible light or whatever lies beyond it – is this individual variation taken into account? If any standard is set, or product approved, based on some average or conception of "normal", what happens to those outside of that range? What provision is there for these people to light their homes or live their lives?

For surely, if light affects nearly every cell in our body, and every physiological system from bowel movements to blood sugar regulation, shouldn't it be subjected to testing systems at least as rigorous as those for medical and pharmaceutical products? And surely we should pay more attention to how we light our built environment – particularly in those places where people have no choice about the light they are exposed to – in schools, hospitals, workplaces, prisons. Especially when legislation takes that choice away from the general population and renders normal working people disabled and prisoners in their own homes.

While we await the required "further research", I think we can listen to our bodies and trust our instincts. When I first had trouble with light and tried to explain this to others, the most common response was an emphatic "weird". How can you be allergic to something that is so everyday and therefore so innocuous? And I admit I often prefaced my explanations with "This is a weird one, but . . ." by way of preparing the ground, asking the listener to stretch a bit, be receptive to something they may not have heard about before.

But I've come round to thinking it's not so weird at all, and certainly not unknown. Light affects us all. A light bulb may be a normal household product, but light itself is extraordinarily powerful. On top of what is known scientifically, we all know its effect on body and mind both intuitively and experientially. We know light can burn and blind and heal and frighten, enchant, intimidate, intrigue and uplift. We know it can evoke elation, depression, anxiety, wonder and sleep. We know that if you want to seduce someone, you look to candles and soft, warm lighting, and you don't switch fluorescent light on. At the other extreme, we know that light has been used as an instrument throughout the gruesome history of torture, to inflict physiological and psychological damage and to break people down.

"So what's it called, this weird condition you suffer from?" people used to ask. "Is it genetic? Is it contagious?"

It isn't really a condition, I'd mumble; *it hasn't got a name. People have different problems for different reasons. They just call us "light-sensitive".*

Yes, I am light-sensitive. And so are we all. Just like the rest of life on Earth.

5

In the Natural World

Light wakes the world gently at this latitude. The sky softens as it ever so slowly brightens, easing from a rich, deep blue into a curious green. The stars dim gradually – like music faded so skilfully that you feel a pang for its passing but listen all the more attentively as it quietens.

Back home I'm not one for prising my sleepy body from my quilt a single moment before I must, and so the magic of dawn too often passes me by. But here – in Glen Affric in the Scottish Highlands – I wake in the darkness, and, momentarily disorientated, I reach out to feel for the campervan curtains. Then gasp out loud at the clarity and density of the stars.

I'm not used to real dark anymore – the sky around my home on a slope of central Scotland is stained from the lights of industry and the sprawl of nearby towns – so the darkness that greeted me when I arrived at this Highland loch last night was unnerving, but welcoming too. It was cloudy, and I realised – or remembered – that real dark is sometimes described as "velvet"; there is that texture to it, soft and alluring. But now the clouds have dispersed and that dark is just a canvas, a backdrop to the exhilarating star-scape that adorns it. More vocabulary takes on re-freshed meaning: "a pristine sky". That's exactly right – it's clean, as if the stars have been polished. And another one: "gaze" – a sense of awe enshrined in the language; we watch birds and even whales, but we *gaze* at stars.

I get up and venture out into the darkness, step by step, the silhouettes of trees splintering the star-bright sky, and the sound of Loch Affric lapping gently at the rocky shore somewhere nearby.

And that's when daylight starts to soften the sky, green-ing the blue way over there, edging a dimmer switch up ever so slightly until the stars are gone and the colours intensify in the autumn-tinged landscape all around me. The world around me is a forever of wooded mountains bathed in morning mist that lies in downy streaks among the trees. Lower down there are dark pines, and the soft yellows of birch just turning; then ghost trees stand behind them in black and white, like pencil drawings gently smudged, and above them is a layer of sky. Then higher still, improbably high: a ridge of pines up in the sky like some wooded celes-tial kingdom that's momentarily revealed.

Then the sun comes pouring in and the mist scarpers, leaving just a few drifty wisps on the water's surface. Sun-light gilds the Scots pines on the shores and they gleam red against the dark forest behind, their reflections stretching into the morning loch. Close by, a robin's call draws my eyes to the silver birch beside me, its September leaves hanging in soft, yellow diamonds. The feathery bracken at my feet is glowing, backlit.

Ha. This trip was supposed to be a break from my light obsession. But here in this enchanted place, I can *feel* the light addressing me from all around. It's light that tells the green birch leaves to fade to soft ochre, tells the robin which song to sing, that touches these Scottish mountains with the colours of the spice rack: saffron and cinnamon and turmeric.

Far from having a break I'm feeling more aware than ever of the interplay between light and life. It's regaling me with a bigger story. Once upon a time, it's telling me, there were dark nights and light days. And that once upon a time lasted a very long time – a good three billion years – and life on Earth emerged, following a constant rhythm, tapping its collective toes to a steady beat: dawn-day-dusk-night, dawn-day-dusk-night. The length of the beats stretching and tightening gently in response to the seasons and the lunar cycle. Climates changed, continents drifted, mountains rose and oceans swirled and still Nature kept the beat: dawn-day-dusk-night. On and on and on.

And pretty much every living thing developed and evolved and adapted to this rhythm. If the world's a stage, the show is an elaborate dance created by Evolution, the flamboyant choreographer, in close collaboration with the Cosmos as virtuoso lighting designer. They're an inseparable partnership – every footstep of the dance takes its cue from the changing light. It's performed by an astonishing cast of creatures, all flying, swimming, hiding, hunting, resting, growing and blossoming to the beat with breathtaking synchronicity. They're all poised for their prompts:

flowers ready to open their petals, frogs to croak a love song, birds to gather and fly south. Migration, predation, pollination and reproduction all respond to the intensity of the light.

And these cues come from the subtle lights of the night as well as the sun in the day. "Real" night, pristine and unpolluted, isn't dark but a perpetual shifting, glowing light show put on by the moon, stars, planets and celestial activity. Many species of migratory birds become skilled astronomers, able to recognise the patterns of the stars and to adjust their navigation to the tilt and rotation of the sky. Dung beetles use the glow of the Milky Way to orientate themselves, ensuring that even on a moonless night they can roll their precious ball of dung in a straight line away from the competition of the dung heap. Hundreds of species of coral all spawn simultaneously after the full moon, because a gene in their DNA enables them to detect light with enough precision to track the phases of the moon.

Whenever I talk about light, I'm also talking about darkness. They're part of the same whole and I feel we're inclined to underestimate the importance of darkness, just as much as we do light. There's a glib presumption that the world goes to sleep at night because we do. Yet much of life on Earth is nocturnal, and darkness is essential, not just for the sleep and health and well-being of diurnal creatures, but for the rest of life that needs the dark to live and thrive, to travel and hunt and hide in.

Looking for otters tuned me in to this otherworld, showed me that the growing dark heralds not a shutdown

Sunlight: nature's miracle, powering all life on Earth.
"I'd still choose the light of a Scottish coastline if I could."

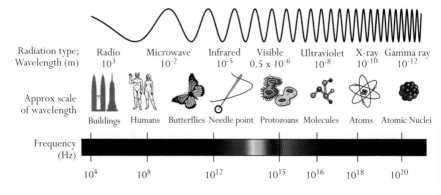

Radiation type; Wavelength (m)

Radio	Microwave	Infrared	Visible	Ultraviolet	X-ray	Gamma ray
10^3	10^{-2}	10^{-5}	0.5×10^{-6}	10^{-8}	10^{-10}	10^{-12}

Approx scale of wavelength

Buildings	Humans	Butterflies	Needle point	Protozoans	Molecules	Atoms	Atomic Nuclei

Frequency (Hz)

$10^4 \qquad 10^8 \qquad 10^{12} \qquad 10^{15} \quad 10^{16} \quad 10^{18} \quad 10^{20}$

Above: The electromagnetic spectrum (Chapter 3), showing the relatively narrow section in the central range that is light visible to the typical human eye.

Below: The 'spectral distribution' (Chapter 3) of sunlight, compared with representative profiles of the wavelengths emitted by LED, incandescent and CFL bulbs.

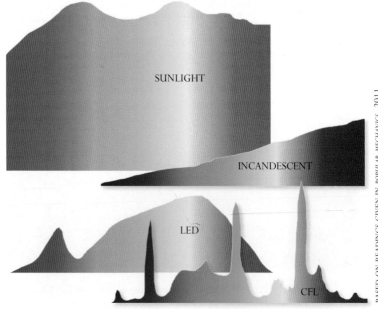

SUNLIGHT

INCANDESCENT

LED

CFL

BASED ON READINGS GIVEN IN *POPULAR MECHANICS*, 2011

Artificial light of different colour temperatures.
L to R: CFL, incandescent, and two warmer-effect CFLs.

Above: Low-pressure sodium lamps in stages of warming up.

L to R: an LED, an incandescent bulb, and a CFL.

Decorative lighting in urban design can make for beautiful streetscapes – but the amount of artificial light at night is escalating unchecked.

M-81 and M-82 galaxies in a starry background (through 14" telescope). The ability to see starry skies is becoming a rare privilege, confined to those who are sufficiently distant from cities.

US States Marine Corps

The Tribute in Light, an annual installation marking the site of the Twin Towers in Manhattan, can be seen for up to 60 miles. It has been used to study the effect of artificial light on migrating birds.

Artificial light in the Americas, from the International Space Station.

The Iberian Peninsula from space, with artificial light
defining the outline of the densely populated coastline.

The Space Station image across Europe, Africa, Asia and Australasia.

Night fishing boats off Taiwan, from the International Space Station.
Blue light is used to attract plankton, drawing squid and fish.

"The history of art is a history of looking at light." Monet portrayed
the changing conditions with 'impressions' of observed colours.

but a shift change. A quiet bridge over a river is a perfect vantage point to observe this process. As the light fades, fishermen and dog walkers head home – the bankside footpaths belong to the roe deer now. The colours drain away and the landscape quietens. A fox might move among the shadows, or bats flicker over the water from riverside trees, and from somewhere in the nearby woodland a tawny owl calls. And sometimes, just sometimes, there's a rippling in the water, a twist and a tapered tail as an otter slips into the river from the darkness of the edges.

Dusk and dawn are when we may notice the light change and feel its impact, but the colour and quality of natural light is changing constantly throughout the day. The shifts in frequency of the light spectrum, from blue morning to red evening, are subtle and often beyond our awareness, but they are not beyond our perception. The cues from lighting designer to cast keep coming, the messages changing: rest now, start to dig, become sexually aroused . . . The performers respond at every level: within the very cells of individual dancers; in their presence on the stage; and in the twists and turns and drama of the interaction between species. If we mess with these cues, what happens to the dance?

And mess with them we have. Big time. Especially in the so-called developed world. Over the past century there has been an exponential rise in the amount of artificial light, and in the brightness of that light. Light pollution encompasses both the direct glare of artificial lighting and sky glow – light that is scattered into the sky and reflected back to the Earth, obscuring the darkness beyond, and the starlight and moonlight within that darkness. In 2016, an international team of scientists measured the brightness of artificial light at night around the world and revealed the scale of the loss of darkness, and warned of negative consequences for "flora, fauna and human well-being". The team reported that eighty-three per cent of the world's population lives under light-polluted skies, and a third of humanity can no longer see the Milky Way due to sky glow, which they describe as "luminous fog".

And so, after some three billion years of constant rhythm, we have abruptly interrupted one of the most fundamental cues that programmes the natural world. What exactly have we done to the dance? We don't yet truly know, so complex are the steps, so interwoven the cast and so subtle are the messages. We've messed up the cues in two ways: firstly with the sheer amount of artificial light at night and the ever increasing extent of light pollution; secondly, and more recently, with the changing quality of that light as a new generation of artificial lighting emits different frequencies of the light spectrum.

Light and the marine environment

The effect of the *quantity* of artificial light is better understood than the *quality* – ecologists have long since raised the alarm about the impact on particular species. Take turtles, one of the best-documented examples. Sea turtles hatch at night on tropical beaches, their mothers having come ashore to bury their eggs in warm, dry sand above the tideline – some having travelled more than 1,000 miles to return to the beach where they hatched. When the hatchlings dig their way to the surface, instinct propels their tiny flippers in the direction of light. Evolution offered them the gleam of starlight and moonshine on the ocean's surface as a reference point to find their way to the sea. And all was well when nesting females could find dark, quiet beaches to gather in large numbers. But humans are also drawn to large, sandy, tropical beaches, and continuous coastal development has created a blaze of artificial light along their shorelines. This can discourage females from nesting, re-routing them to less suitable breeding grounds, or even depositing eggs in the ocean if repeated attempts to nest on land are thwarted. Where females do nest anyway, the hatchlings emerge to find light coming from the wrong side of the beach. Light from houses, hotels, nightclubs and cars lure the tiny turtles away from the ocean and into the dangers of busy streets and human habitation.

In many regions around the world, attempts are being made to mitigate the effects of light on turtles – there are projects to dim lights, shield beaches during the breeding season and teams of volunteers to pick up disorientated turtles and return them to the sea. But on a global level, light at night remains a big problem for sea turtles, which already have trouble enough: direct hunting, entanglement and marine pollution have resulted in all seven species of sea turtle being endangered.

The satellite images of Earth show light clustered

around tropical shores where turtles need darkness. The lights shape the map, defining the coastline with a jewelled outline of white and amber crystals, shining, set against a rich, black edge of ocean. But further out into that black there are more lights, jewels scattered into the darkness. Yes, even in the ocean. You'd think – or at least I'd thought – that the seas might be one last sanctuary of darkness in this world. In more desperate moments I've dreamed of gathering the scattered clan of light-sensitive people and escaping together somewhere. Harsh light is encroaching, closing in like an enemy army and we need to flee. But where could we go? Sometimes I've dreamt of an island, or even a ship – sailing away with the last stash of incandescent light bulbs on board and starry skies above. But now I find that artificial light is one more pollutant to add to litter, chemicals and sewage in the almighty mess we're making of the ocean. As well as a blur of sky glow from the coast, there are direct lights from ships, oil rigs and fishing vessels.

In 2012, NASA released data of the Earth at night using new and more sophisticated techniques to observe light. Among the surprises it revealed were exceptionally bright lights in the ocean off the coast of South America and in areas of Southeast Asia. Lights so bright they could be tracked from space, and which have led people camping along the Argentinian coast to report UFO sightings on the horizon. The satellites showed that the lights are clustered around the edge of the continental shelf, where ocean currents meet and create a rich soup of nutrients. Here an explosion of phytoplankton (tiny marine algae) draws in fish and large numbers of squid, which are followed by squid fisheries. Nearly ninety per cent of the world's catch of squid is caught using light. Known as "jigging", the fleet uses powerful lamps to draw fish and squid from the depths to the shallows, where they're caught on a jigging line. There may be more than a hundred bright lamps on board each boat, constantly blasting the sea with intense light.

Fishermen have always known that many species of fish are attracted to light, and have long exploited this phenomenon as a fishing strategy. For thousands of years fires were lit on beaches, or torches made from plant material were carried into the water. Then fishing boats set out with gas lamps, then electric lights, then ever-brighter lights: fluorescent, LED and metal halide are all used in commercial fisheries today.

What are the implications of this level of light in the ocean? As ever, it is easier to know what is *attracted* by light than what is *repelled* by it – or what unknown consequences ripple on through the food chain. We know that many marine organisms are extremely sensitive to light. US oceanographer Edith Widder talks of the "language of light" in the ocean. She has studied bioluminescence for more than thirty years, since becoming enchanted by the "underwater fireworks" on a dive. She explains that most deep-sea organisms use light to communicate: the cookie-cutter shark has a bioluminescent stomach to disguise its own shadow; one species of dragonfish can emit and perceive red light; and there's an octopus with suckers that have evolved into lamps. What are we doing to that language if we blast the underwater world with artificial light? What knock-on effects does this disruption of communication have on energy, breeding success, predation and prey?

Artificial lights in the night can also be a danger for seabirds, especially nocturnally active species that feed on bioluminescent sea creatures and are alert to low levels of light. Other species are drawn to the direct lights of fishing fleets, which can distract them from their migratory course or lead to collisions with vessels. Birdlife International, a global partnership of conservation organisations, recognises light pollution as one of the threats to birds, especially seabirds. They draw attention to the "immense scale" of the problem of seabird attraction to artificial light, documenting one incident when 6,000 crested auklets nearly capsized a fishing fleet when they were drawn to its lights and landed on board. On another occasion in the South Georgia area of the Atlantic, 900 seabirds, mostly petrels, collided with a research vessel using powerful searchlights to spot floating ice.

It has long been known that seabirds can become "trapped" in beams of floodlight, such as from lighthouses, continuously circling within the beam until they collapse from exhaustion. This situation is unwittingly replicated in New York, where an annual memorial at Ground Zero sends two vertical columns of light into the sky to represent the Twin Towers. Migrating birds are drawn into the floodlight in large numbers: an international study monitored the bird behaviour over seven of these nights and revealed that more than a million birds were affected. Such a specific light has a measurable response, but this phenomenon – of birds drawn to light – is happening globally on a scale beyond measure. Migratory birds use the sky at night as one of their navigational tools and can be disorientated by sky glow, and drawn off course towards cities. Arriving too early or too late can mean being out of step with plantlife, insects, nesting and foraging, and all the complex synchronicity that's required to thrive.

That said, not all effects of light are negative – some species are thriving under artificial light conditions. The oil refinery at Grangemouth in Scotland spills constant light onto the mudflats and inter-tidal areas of the Forth estuary, an important habitat for wading birds. A study on common redshanks revealed that these elegant waders are feeding well there, with the continual light creating a very convenient 24/7 service station, as far as they're concerned. But not all "wins" for a single species work for a whole ecosystem – feeding or breeding success can knock predator and prey relationships out of balance.

Light on the land

And so to insects and other invertebrates – the unsung heroes of biodiversity, with more than 900,000 known species, representing most of life on Earth. Among their myriad forms, and eyes, and mechanisms to detect light, is a diverse range of sensitive relationships to light. Some

– such as moths and midges – are among the best-known examples of wildlife attracted to light. Other invertebrates are repelled by light, including earwigs, cockroaches, wood-lice and earthworms. Fireflies, also known as lightning bugs and glowworms, create their own light by bioluminescence to seduce potential mates. Many insects are diverted by light from where they should be or what they should be doing: reflected light can draw pollinators away from their target crops, and aquatic insects are lured from water by polarised light on shiny surfaces nearby.

A report by UK invertebrate charity Buglife estimates that a third of flying insects attracted to street lights will die as a result, whether from collision with a lamp or increased susceptibility to predation. And light is also the trigger for the developing stages of many invertebrate life cycles, so how will changes in light affect these? As artificial light spreads, and changes in intensity, some insects may be wiped out, while others will thrive. The critical issue is that invertebrates are so fundamental to the functioning of the wider ecosystem that any imbalance in populations can have magnified consequences. In 2019, a global review warned that the rate of insect extinction is eight times faster than that of mammals, birds and reptiles, and that the current rate of decline threatens "a collapse of nature's ecosystems".

Insects attracted to artificial lights draw bats in pursuit, and studies of this interrelationship were among the first to highlight that the quality of artificial light is changing, as well as the quantity, and to question the impact of this. One study followed the lesser horseshoe bat, a shy, slow-flying species that stays close to home – normally foraging within a mile of its roost. It found that light at night dramatically reduces the bats' activity and delays the timing of going out to feed. Further research has indicated that there are varied responses to different types of light: there is significantly more hunting activity by bats under white metal halide street lights than under orange sodium lights.

The effect of light on different insect species has a knock-on effect on bats. In response to bats' echolocation calls, some moths perform a "last ditch evasive manoeuvre" – a sudden power dive to dodge the bat's flight path. But they do this less under LED lighting, tipping the balance of the predator–prey relationship in favour of the bats. A study of insects under different street lights showed that white lighting attracts more insects, with metal halide having the biggest impact, while the greatest diversity of insects is found under LED. Together, such studies demonstrate an important and complex relationship between light, bats and insects. As street lights are changing fast throughout the world, we need to understand more about the different properties of light used and the effect on wildlife.

A recent project in the Netherlands lit an otherwise dark and undisturbed habitat with lamps emitting white, green and red light. They found that pipistrelle bats, which are agile and opportunistic feeders, are attracted to all light, probably because of the insects gathered there. However, mouse-eared bats and long-eared bats, both forest-dwelling species that fly slower, avoid the white and green light. The project concluded that under red light, the bats displayed the most natural behaviour – closest to the way they behave in the dark. Therefore, using red light – rather than white or green – would be best to minimise disturbance when using artificial lighting near bat habitat.

An ecological lightmare

The new generation of lighting tends to use more of the blue end of the light spectrum, especially for outdoor use. If this blue-rich light has such an effect on human minds and bodies, what is it doing to the rest of the natural world? NASA satellites can't detect some of the blue frequency, and so the true picture of night-time light is even more extensive than the current data shows. In general, the way we document, map and measure light is based on the way humans perceive it. But what do other forms of life experience? This varies dramatically between species and even within different life stages of the same species: moths and butterflies from their caterpillars, for example.

The Environment and Sustainability Institute at the University of Exeter is a hub of investigation into artificial light and the natural world. The research is led by Professor Kevin Gaston, who tells me he has seen an "explosion of interest" within the natural sciences as biologists awaken to the scale of the issue, and the questions within questions that it contains. The growth in the extent and intensity of light, as well as the changing spectrum of light, adds up to a "perfect storm" of ecological impact.

"This is an issue whose time has come," he says. "Every few days new papers arrive on my desk. There's a realisation that it is a lot more extensive than people had envisaged, and this is not just an urban issue."

Ultimately, he explains, changing light means altering the perception of time in the natural world, something that is "extremely novel in evolutionary terms": "Organisms have experienced many pressures over geological time, but what they haven't seen is us messing around with their ability to tell how long a day is."

This has a profound impact on everything, including plants, which are showing shifts in growth patterns and flowering times. Disrupt the timetable of the natural world, and other species can miss their connection. If plants are changing the timing of development, how does that synch with other species that depend on these plants, such as caterpillars emerging or migratory birds arriving? There is the potential for amplification up the food chain, depending on the ability of different species to track and adapt to these shifts.

"What we need overall is a much more nuanced sense about lighting," Professor Gaston says. "It's not appropriate to deal with street lighting as a 'one size fits all'. We need to ask what the lighting is for – is this a protected nature area, or a major vehicle junction, or are there concerns about sleep in a residential setting?"

While the science is firming up fast and growing our understanding of the ecological impact of artificial light, there is still a big gap between ecology and policy. "Lighting engineers are developing products around regulations whose scientific grounding is open to question," says Professor Gaston. "I don't think environmental impact ever entered the debate.

"It was hoped that the rise in LED lighting would reduce energy use and artificial light at night – but both are still increasing," he says. "Between 2012 and 2016, the amount of the planet artificially lit at night grew by more than two per cent each year." He describes the "rebound effect" in which the lower cost, and the perception that LED lighting is energy efficient, actually increases the use of light.

"Many environmental challenges today are difficult and expensive to solve, from ocean acidification to climate change. The oddity about light pollution is that you could change it very quickly – with huge economic savings and very little evidence of a downside – simply by reducing the quantity of artificial light at night.

"There are really interesting issues around the psychology of night-time lighting – how rapidly Western society has formed norms and expectations of light at night. The spread of artificial light is a relatively recent phenomenon, and we have come to accept this so fast and to equate more lighting with progress."

It's seen in simplistic terms, he feels, of "light is good, dark is bad": "Often there isn't a real, tangible benefit of artificial light at night, but only to alleviate people's fears. It taps into something fundamental – where fear is unacceptable and anything we're afraid of is demonised and dealt with."

The case for darkness

My quest to understand more leads me to Dumfries and Galloway, a gentle corner of southwest Scotland, for the first European Dark Sky Places Conference. It takes place within Galloway Forest Dark Sky Park, the UK's first area designated as such and keen to celebrate its status and pristine skies. I find myself among a delightfully multi-disciplinary global gathering: academics, astronomers, artists, tourism people, rangers and guides, neurologists, social geographers, street light engineers and more. And they have all come to talk about light! And darkness, of course. Darkness and light. Having reached the stage where most of my friends and family flinch or glaze over when I talk about light, it is a joy and relief to do so all day long and through most of the night. I meet with the owners of a "cosmic campsite" in Arizona, hear of a pitch-black restaurant in London, and learn about Sark, one of the Channel Islands, where there are no cars or street lights and the people are not afraid of the dark.

At breakfast I find myself sitting beside Dr Thomas Davies, a marine ecologist from the University of Exeter, who argues that the environmental impact of changing light is *as significant as a changing climate*, but far less recognised. He works on the ecological effects of different spectral compositions of LED light. In one project, the team exposed a group of dog whelks, a type of sea snail, to bright white LED street lighting, while a control group were kept under a natural cycle of day and night. The dog whelks under artificial light became "metabolically stressed", with their foraging and risk perception affected as they spent longer looking for food and were less likely to seek shelter. Dog whelks are widespread around rocky UK coastlines and play an important role in the ecological balance.

The problem as Dr Davies sees it is not with LED lighting *per se* but with inappropriate and insensitive use of overly bright light. He feels that LED technology could be a potential solution, as well as a current problem, as the lights can be created in such a range of spectral composition that it may one day be possible to develop appropriate lighting for different habitats and ecological niches. There is growing interest from lighting manufacturers and policymakers in "ecologically appropriate lighting", but for this to be realised would require great cooperation and understanding between engineers and ecologists, and far more long-term research.

For here's the catch: if we use "spectral manipulation" – changing the colour – of outdoor lights to avoid impacts on some species, others may still be adversely affected because there are so many different biological responses, so many unanswered questions. The different responses of different bat species to light is just one example.

Ultimately, Dr Davies concludes, we need to question whether we need so much artificial light at night in the first place. Can we overcome our cultural fear and negative associations with darkness? Can we call on our artists, poets, astronomers and night rangers to help us appreciate and celebrate dark skies? With more than eighty per cent of the world population now living under light-polluted skies, we're facing an "extinction of experience" of true darkness. Does this perpetuate disengagement with the environment, and does this lead to a lack of advocacy for the natural world?

Which brings us back to humans – our beliefs, perceptions and values as drivers of this global environmental change. Where do people fit into the ecology of this?

"I think there's a disconnect between those developing the lighting and the end users," Dr Davies says. "There's a perception that everyone likes the new LED street lights, but I'm regularly contacted by people who hate them and are looking for ecological data to bolster the case against them. Personally, I find white LED lighting when driving extremely hazardous, affecting long-distance hazard perception. The contrast is too great: it's like repeatedly coming out of the cinema, blinking in the light.

"We don't have the information to make educated decisions about light. I draw parallels with cigarettes: once, doctors would even prescribe cigarettes for certain conditions; now, you can't buy a packet without gruesome pictures of damaged lungs on the front. The public don't yet know about light – what it does to our bodies and minds – and so cannot make an informed purchase."

Dr Davies grew up by the coast, "intuitively appreciating

the natural rhythms, reading the tide and the weather patterns – it takes you to a place where you fully integrate with the environment". There's a lot of talk about "connection" at the Dark Skies Conference . . . connection to the Earth and to the stars. For it's not just about the birds and beasts – what do *we* lose when we lose sight of the night sky, many delegates are asking. And the same answer keeps coming in different forms: a connection with something greater than ourselves. Our universe.

A bright blue sky has appeared outside while we've been in meeting rooms, talking and talking about darkness and light. I feel a sudden need to walk outside alone to think things through. I stroll around the expansive grounds, sweet September light casting long shadows across the hotel's lawns. The wet grass gleams, reminding me that light fuels the natural world through the alchemy of photosynthesis. It's by light that leaves live, and plants grow and feed animals, and so sustain the intricate food chains that bind together all the flora and fauna on the planet.

For someone pretty alert to the world around me, and working in wildlife journalism, I was surprisingly slow to pick up on the extent to which this changing light was a natural history story. I think I'd been so bewildered by my own struggles, so troubled by others' pain, and so intrigued and confused by the legislation, that I'd been viewing it all through a narrow lens as a human story, about power and politics, values and justice. But in the sweet evening light of Dumfries and Galloway, I realise that what I've learnt about so far – how this stuff of physics impacts on skin and eyes and nervous systems and hormones and sleep and temperament . . . we share these things with the rest of creation. We're all programmed by the same cues. Dog whelks are getting metabolically stressed by blue-rich LED street lighting, and so are we.

When I've given talks to schoolchildren, the only way I've found to explain "ecology" is "the connected-up-y-ness of life". And now I feel a delicious sense of the connected-up-y-ness of so many disciplines – but also of my own life and work. In getting embroiled in this story of light, I had felt I was being tugged away from my natural habitat of wildlife writing into a deeper, murkier world where I found myself facing hostility to my stance on the lighting ban from much of the scene that I'd previously felt a part of. But now things have come full circle and I find myself right back in the wildlife world again. For this is a story about my life and Jesse's and Elaine's, and about dog whelks and seabirds, birch leaves and songbirds, fireflies and phyto-plankton, the moon and stars and Milky Way . . . it's about no less than all of everything.

6

The World We've Created

This feels good. Steam from my tea is curling gently into a beam of afternoon sunlight that's streaming in from the window behind me. The lights are on – those "zigzag" ones we're seeing more of in fancy cafés – and they give the place a golden glow. It's all good – the staff are smiley, the music's lively enough to uplift but gentle enough to let me think straight, a slice of cake waits as my reward, and even the blank white page of my notebook looks inviting not daunting now.

I've always preferred working in cafés, rather than at home where the house is teeming with domestic demands. Most libraries are out of bounds now because of the lights, and it's getting ever harder to find places to go. So I had been wandering the streets of Falkirk, searching for a café to write in. I looked inside each coffee shop or restaurant, scanning for inviting lighting the way other people may peruse menus. As I made my way around town, I felt attraction and repulsion on a cellular level. A real fire beckoned, fairy lights called me enticingly, but naked CFLs or big fluorescent strip lights made me wince and flinch away. How can you *eat* in there, I wondered, passing a

138

fluorescent-lit café where a guy was tucking into an all-day breakfast. The blue note in the light met the pale pink of his bacon and my stomach recoiled.

Those zigzag lights are known as "squirrel cage" – they are incandescent wire lamps, which have now been banned and will soon be gone. There are LED ones designed to look very similar, but peer closely and you see they're made of small flat strips, not wires. I find their light warm, too, but with a harsher edge. I look around the café at people drinking coffee, reading papers, a couple chatting. When the stockpiled incandescent bulbs have all gone, *colours* will change. Cafés will look different. As will sitting rooms and streets throughout the land.

I've been thinking about artificial light and the environment as being about the natural world: trees and bats and turtles and seabirds and all. But of course there is also the environment we have created around us – our houses and cities and streets and buildings. How do changes in artificial lighting changing our immediate surroundings, and who notices or cares? Does this only matter to those of us who feel immediate negative effects from new forms of lighting, or is this something more universal and fundamental?

In the years that I've been talking about light, relentlessly, I sense there are two sets of people: those who "get it", instinctively – they notice light and know it matters; and those who just frown and blink at me as if it's something they've never considered before, or only thought of in terms of figures or physics. It's as if some people have an extra perception, a radar of light awareness lodged somewhere in their soul.

The art of lighting

"Ideas are to literature what light is to painting."
– Paul Bourget

Art gets it, as a rule. Artists of many kinds talk eloquently and often about light. How could it be otherwise? Art and light are intertwined in so many different dimensions: the work itself, its display and its longevity. Photography is the very art of capturing light, rendering it in a form we can keep. And those who express their art in the creation of the human environment – in landscape and architecture – they talk about light, too. "More and more, so it seems to me, light is the beautifier of the building," said US architect and writer Frank Lloyd Wright, spearheading a renewed appreciation and awareness of daylight.

Light is everything in art: the direction the light comes from; the colours it creates and the forms it reveals; what is highlighted and what is softened in so many layers of shadows. Once, I showed my young son that black is not just black – there are infinite shades and nuances. He was doing some colouring in at the table and we stopped and looked at our black dog curled in his basket, snoring softly, and a black fiddle case propped up against the wall. He glanced, then really looked, and I saw his face change as it clicked. There was a light dusting of copper where the sunlight fell on the dog's flanks, rich blackness in the folds of fur around the neck, a silvery triangle on the ears – the same shade on one slope of the fiddle case. His colouring in changed from that point on.

We visit an exhibition at the Scottish National Portrait Gallery, and while the children hurtle around – "See this one!" "No way, that's got to be a photo!" "Look at her *hair*!" – I'm taking in the levels of light involved in our experience. There's the exhilarating range of media and techniques for depicting the way light falls on faces. Then there's the way these photographs and paintings themselves are lit – the spotlights and strips and ambient, diffused light. A German gallery owner has described the ban on incandescent light as an "aesthetic calamity", but LEDs have advanced so far that there now seems to be widespread enthusiasm for the flexibility and possibilities of LED for lighting art galleries. In an age where accessibility and inclusion is such a prominent part of arts discourse, does anyone know that some people cannot tolerate LED lighting and so will be excluded from enjoying the nation's treasures?

I meet my aunt, Cathy, for lunch and grumble, as I so often do, that people don't talk about light. But Cathy certainly does: she has been caring for museum collections for forty years. Light has been a constant focus in her work: a frustration, inspiration and bone of contention. Those responsible for curating collections have to reconcile light in all they do – how the light chosen alters our view of the work, how true that is to the original creation. There can be tension, she tells me, between the visitor experience and the preservation of collections because light also interacts physically with the work itself as an agent of natural degradation.

When Cathy began work as a conservator, her involvement with lighting was more to ensure that brightness levels, measured in lux, were under control – certain pigments in paint and many natural materials are particularly affected by light, fading faster than others and so altering the appearance of the work or object and also threatening its longevity. Watercolours, for example, are particularly susceptible to light and the standards for light levels were set at no more than fifty lux – "which would only just be okay to someone with good eyesight, and no use for anyone over the age of forty".

But Cathy found she began to recognise the importance of good lighting, not just for the preservation of the collections, but as what enables people to be able to enjoy them and see them properly. She has encountered plenty of bad lighting in her time – absurdly low light levels, dazzling reflections, spotlights shining in people's eyes, black holes in displays, distracting flicker. And she's seen fashions come and go: for lots of daylight, or none; for directional spots or even wash lights; for the atmosphere of a retail display or an operating theatre.

Nowadays she feels there is more understanding and flexibility about light levels, but anxiety about light damage can still spoil people's experience of a museum display. Many small galleries are at the mercy of generalist electricians who may not appreciate the subtle nuances of lighting a specialist environment, or sales reps masquerading as lighting designers to promote their products. What's needed, Cathy says, are specialist lighting designers as part of a display team, who understand the challenge of the limits set by conservation. And there needs to be mutual respect and understanding from the rest of the team that effective lighting is crucial to the visitors' access to and enjoyment of the collections.

Lighting and design

> "Oh joy! There was an archway filled with a quite different sort of light; the honest, yellowish, warm light of such a lamp as humans use."
>
> – CS Lewis, *The Silver Chair*

Lighting design is under-recognised as an art form. Lighting designers use light the way composers use sound or artists use colour – it's their palette and their medium, to create and conjure with. It's no surprise that lighting designers were quick to raise the alarm about the legislation banning incandescent lighting (or "tungsten", as theatre people usually call it). At the first stage of the EU ban, lighting designers in the UK mobilised a "Save Tungsten Campaign", which described the importance of incandescent lighting in their creative world:

> Stage lighting is telling stories with light . . . We must have flexible light; light that can be warm or cool; light that can be ever-so dim or blindingly bright, light that

can subtly or brashly change its characteristics, full wave-length light that can truly reveal every colour in the spectrum; but most of all, beautifully illuminate the human face, the humanity, which lies at the heart of all our work. At the present time, that means that the incandescent lamp is an essential weapon in the designers' arsenal . . . it cannot be done without.

After years of isolation, I feel a sense of relief and reso-nance flooding through me when I read these testimonies. When people get it. When they know: that light matters, and that the quality of light affects the quality of our lives, our every waking moment – and every sleeping moment, too, for that matter. I wonder: do people become lighting designers because they know this instinctively, or does the process of designing lighting create such awareness?

"A bit of both," says Kevan Shaw, one of Scotland's most prominent lighting designers, who meets me in one of Edinburgh's last remaining incandescent-lit pubs in the shore district of Leith. He's softly spoken with a wry smile, understated and thoughtful. Kevan's father was a stained glass artist and he spent many childhood holidays being "dragged around French cathedrals". Maybe it "set some-thing going subliminally about the power of colour and light," he says.

Kevan drifted into lighting through his interest in the-atre, then working with lighting rigs as a roadie for rock and roll bands. This ignited a passion for lighting design as he found he had strong feelings about light and colour, and he experienced the buzz of eliciting a reaction from a crowd – that by manipulating light you could "get an *ooooooh* from thousands of people!" He went on to create special effects with a theatre lighting company in London, including ambitious installations such as enormous water features and the effect of a volcano erupting outside a Las Vegas hotel. Returning to Scotland, he set up his company Kevan Shaw Lighting Design in 1989, initially specialising in historical buildings, museums and exhibitions.

People experiencing lit spaces, he says, are often unaware of the light and how it is affecting them – whether in the theatre or beyond. "A lot of what lighting designers do is subliminal – good lighting is to a large extent unobtrusive. People may look better, feel more energy, but they don't necessarily associate that with the light. Awards for lighting are often about big, flashy stuff – but good lighting doesn't necessarily stand out or photograph well: you just feel something, experience something, in that space. Lighting is just the tool and unless there's a specific reason in the theatre experience, you shouldn't see the way it works. Because people who are not engaged with lighting have no clue of the impact of light on them, you've almost got to do something bad to attract their attention!"

Kevan confesses to spending time hanging around the light bulb section of DIY shops watching how people make lighting decisions. It's almost always about the initial cost, he's noticed. But there should be much more to it than that. "I've always been aware of the impact of health and well-being and what you can do with light," he says.

"Coming from a theatre background, you're communicating so much: time passing, mood and atmosphere. You can move huge chunks of understanding around with light. If you do that on a stage, think about what effect light is having on a wider environment.

"We need to understand the root cause of health problems with new lighting," he agrees. "We have so much to learn about light. Vision is a lot less mechanical than it is portrayed – it's all in the mind not the eyes. For example, lucid dreaming is a full visual experience. The explanations for the optics of colour are inadequate for the complex communication from eye to brain. Stevie Wonder was profoundly blind but totally aware of the light shining on him – he could name the colour and was very specific in his requirements for lighting design."

Always interested in innovation, Kevan was an early adopter of LED lighting and excited to explore the potential of this new technology. They are brilliant for some applications, he says, but the rush to legislate meant much equipment was introduced far too quickly, rendering "millions of light fittings obsolete though they still worked perfectly well" and forcing LEDs into forms and shapes that were "grossly unsuitable".

As a lighting designer, what do you lose, I ask, when the incandescent bulbs disappear from your toolkit? I'm slightly braced for his reply, anticipating complex technical explanations beyond my grasp, but he says just one word: "Warmth." When pushed to elaborate, he adds "softness" and "naturalness". And that he feels the impact most in restaurant design, or "anywhere you eat". But he's already said it all.

Kevan returns to the demands of his busy office, and I take the chance to explore Leith's vibrant docklands, feeling the cold wind rolling in from the North Sea. I pass a pub with inviting lighting and a blazing fire inside, and I decide to stay a while and write up my notes. I settle as close as is physically possible to the fire and can feel the heat seeping into me, through my jeans and on my face. My hands are glowing, gilded with warmth. ". . . anywhere you eat," I'm typing up. *And anywhere you take your clothes off*, I had been thinking at the time but didn't say. Where I notice light most is on skin: whether in a restaurant or the bedroom, light changes the colour of skin. And it's something more than colour – it uglifies or beautifies.

The UK's leading lighting designer for dance is a passionate proponent of incandescent lighting. Michael Hulls is described as a "choreographer of light". He's an associate of London's Sadler's Wells Theatre and a recipient of an Olivier Award for Outstanding Achievement in Dance.

Michael became a lighting designer by accident: he actually studied dance and theatre, but one day a friend asked him to do the lights for a dance, and then it clicked. "I improvised lighting while they improvised dancing, and I loved it," he says. This interest in improvisation led to a long-standing collaboration with dancer Russell Maliphant, which is still going strong twenty-five years on. "Russell is described as a solo dancer but it's a duet – an interaction with dance and light," Michael explains. "I noticed Russell was very sensitive to light: he'd dance differently in the centre of a pool of light than at the edges of it. Sometimes we make lighting first and he choreographs to go with that."

For Michael, this style of working gives much more freedom to explore and create than working in a conventional theatre setting, where it's necessary to "apply" the light to something that already exists. "If you're lighting a narrative, you're at the service of that narrative. Lighting dance in this way is the greatest freedom for light to achieve its potential and to be whatever it wants to be – what happens then is that the light becomes the narrative. That's the most exciting thing."

Michael first caught my eye with his art installations, *Lightspace* and *Castor and Pollux*, which feature glowing constellations of incandescent light bulbs – they are a celebration of, and requiem for, the tungsten bulb. "There is nothing as beautiful as a piece of tungsten wire glowing in a glass bulb," he says. "Even before I'd agreed to do the installation, I'd filled my garage with hundreds of tungsten lamps. I knew they were going to disappear very soon so I hoovered up the remaining stock I could find. I was ringing suppliers, wholesalers in Europe and manufacturers to find out what was left out there."

How important is this light to your work as a designer?

"In my toolbox, most of the tools I use are tungsten, but I also use video projection, arc sources and LED. There is a huge amount of LED lighting out there and much of it is really poor quality, but the good stuff is getting really good now – there are some really useful tools. Certain LEDs can now be programmed in a way that mimics tungsten."

It's not a replacement though, he says; LED will always be a different thing. One challenge with even the best LEDs is lighting skin tone. A stage set might look fantastic under LED lighting, but when a dancer comes on stage it isn't quite right. "Tungsten gives a truth to skin, you see the real colours," he says. "It's emitting the entire visible spectrum and all wavelengths of light are coming out and bouncing back to your eye. There's a truth to the rendering of colour and, as of yet, even the most sophisticated LED can't do that."

I ask how he feels about the ban on incandescent lighting and his tone changes abruptly.

"It's incredibly important to reduce carbon emissions, but it doesn't actually matter how the fuck you do it – it's nonsense!" he says. "We're being sold a fiction that it's better for the environment. Low-energy lighting might be 'energy efficient' for the hour it's on, but nobody is measuring the whole story, no manufacturer is going to go there. Nobody takes into account the total life cycle and energy consumption, the digging up of rare earth elements – studies show there are toxins including arsenic used to create LEDs. CFLs contain mercury and nobody disposes of them properly, and the quality of light is horrible. We're being forced to light with things that are quite poisonous. Whereas tungsten light is easily made anywhere in the world; it's easily disposed of and recycled."

And all of this applies to the world beyond theatre and

the stage – to people's homes and the outdoor environment. "One of the bad things about LEDs is that you can put them anywhere, so people do put them everywhere. There are horrid blue LEDs under the railway arch near my home, a really disturbing wavelength of blue light and they're on 24/7, 365 days a year. We haven't been in the world of LEDs for long enough for things to be completely apparent. For example, as they wear out their ability to create colour diminishes."

Do we underestimate light?

"We're phototropic creatures, but we don't recognise that and we don't talk about it, yet it's so pervasive, so fundamental."

Michael is, he says, in an understatement that makes me smile, "quite sensitive to light". Certain frequencies feel like nails on a blackboard. Hell would be a locked room with fluorescent strips overhead. But incandescence enchants: "I find it as mesmerising and meditative as staring at an open fire; it takes my mind to the same place."

It's an attitude echoed by American lighting designer Howard Brandston, who had this to say about the US incandescent ban: "The ban will dim everyday life as we know it. CFLs rate well below incandescents on the colour scale, producing less light and poor colour by comparison. Headaches will escalate. Art, your loved ones, and objects will never look the same. Get ready for a general atmosphere of sickly light and gloom . . ."

Howard has talked about light all his working life. He's lit the Statue of Liberty, the Petronas Towers in Kuala

Lumpur and the American Museum of Natural History. He boasts that his company has worked in every continent but Antarctica: "There are a lot of people wandering around in the light I've cast around their environment."

Howard feels he has "an appreciation of the contribution that light can make to space", and this led him to his vociferous opposition to the lighting ban in the US: "My position is that there should be no ban on any lamps, just a market," he states. "I'm not opposed to LED or CFL lighting. I'm opposed to banning a perfectly fine light source, which in over a century of use has had no health problems, no functional problems. I use LEDs for some things – they have some useful contributions to make – and they have helped to advance the art of lighting design, but that doesn't mean they're primary sources. Incandescent light has optical properties as well as spectral properties that LEDs and fluorescents can't match. I don't think they will ever get a perfect replacement for the incandescent lamp. You can't beat it. It's a very simple product, very inexpensive to make."

Lighting the built environment

"A single mercury-vapour lamp that looked like it had illuminated the end of the world more than a few times."   – Haruki Murakami, *1Q84*

Beyond the theatre, have we forgotten that lighting design is an art, or that all the world's a stage? Nowhere is this more apparent than street lighting and other outdoor lighting, which "paints" the colour, tone and atmosphere of our built environment.

In the decade we've lived in our home in central Scotland, the street lights have been changed three times. First the orange sodium glow was replaced with the ghoulish glow of white fluorescent; then the fluorescent was replaced with a neat rectangle of LED light, a pool of light below it and darkness beyond. There was never any notification, consultation or discussion – you'd just see a few guys around a cherry picker truck one day and then the evening would change colour and atmosphere.

It's a microcosm of the changes that have swept the world. The spread of LEDs can be seen from space, as cities glow brighter than suburbs. The first wave of LED street lighting was focused on cost and efficiency, and the most cost-effective and efficient LED lights are blue and bright.

Of course, this has been devastating for those who cannot tolerate LED lighting, but it was the aesthetic shift that caused far more media reaction, as people around the world baulked at the site of familiar landscapes rendered altered, colder, in the new harsh light. This hit hardest in places renowned for the romantic atmosphere of evening townscape. A municipal council member from Rome was quoted in the *New York Times* as saying, "Illumination is atmosphere. [LEDs] are assassins of the beauty of Rome, of its history." Another resident commented on social media that the change was akin to "candlelit dinner versus the frozen food aisle of your local grocery store".

"Cities across Canada are swapping old street lights for more energy-efficient LED models and facing an insurmountable problem: a lot of people hate them," reported CBC Radio in 2017. But not everybody hates them. There is also widespread enthusiasm for the new technology – especially its adaptability and enhanced possibilities for adjustment of lights. Cash-strapped councils throughout the UK have seized an opportunity to get more light for less outlay, saying that there will be huge cost savings in time from big reductions in energy use and maintenance cost. They claim they are hugely reducing carbon emissions and enhancing the safety of residents when they change to LEDs. I met a street light engineer from a Scottish council who was literally bouncing up and down on his feet like an excited toddler as he enthused about the "win-win" of LED street lighting. "It's cut our huge maintenance budget! Bounce, bounce. "It's helping meet our carbon reduction target! People love it!" Bounce. "They say they sleep better, feel safer!"

But as a rule, LED street lights have been rolled out too soon, too bright and too blue. Towns and cities are changing colour, making evenings out a qualitatively different experience, especially when a soft orange glow changes to a harsh blue glare. Some accept it stoically, while others embrace it enthusiastically. Then there are those who blow a fuse. Collectively and individually, in petitions, campaigns, letters to newspapers and lawsuits, people have found ways to express their opposition.

Incandescent

Simon Nicholas is an engineer who lives in a conservation area in Greater Manchester. Elsewhere in the country, he has encountered "shocking" 5700K street lights with "hideous aesthetics". So when he discovered that LED street lights were earmarked for his area, he tried to take the council to court, arguing that LED street lighting is rushed out without proper procedures to assess the risks. His one-man campaign delayed the roll-out of LED street lighting twice, but he was refused a judicial review on a technicality of timing. Undeterred, he has continued to campaign against a lack of scrutiny and due diligence in street light conversion.

LEDs differ from *all* other light sources, he argues, and have certain optical properties that make them fundamentally unsuitable for many tasks. "For street lighting, we need a light source that illuminates uniformly and has no sharp shadows or debilitating glare. LED is an *acutely directional light source*, more akin to a laser beam than a conventional light source. There is a huge increase in light intensity at the centre, excessive glare and a sharp cut-off."

Simon explains that LED lights were initially classified along with lasers by the international standards for assessing potential harm to the eyes. In 2006, LEDs were reclassified into regulations for lamps. "And at that time it was said that we needed a third standard. LEDs are not lasers, but neither are they general light sources. As you move around and view an LED you suddenly encounter extremely high luminous intensity in the central plane. It's like when you look at a laser on a wall – the wall is dark except for the bright circle. LED is also extremely concentrated at its source. I believe chronic, repeated exposure will have a huge impact on ocular health.

There is also a danger to motorists and pedestrians, Simon argues, because LED street lights over-illuminate one spot, with dark patches in between. "I've heard people talk of 'pavement ghosts' – motorists don't see pedestrians stepping out of the shadows. This sudden contrast is a particular problem for the eyesight of older people, and there are further complications from the effect of wet weather and snow."

LEDs were first developed as marker lights and backlighting on electronic devices. "It's a good tool for an area you want to illuminate in close proximity to the source, but it was never meant for illumination of a wide area. Trying to make it do so is to bang a square peg into a round hole."

The tide began to turn. In 2016, the American Medical Association issued an official policy statement about LED street lights, especially aimed at large cities where lights of a colour temperature of 5000K or 6000K were common. The statement suggested that these lights could have adverse effects on health, especially on eyes and sleep. The recommendations were for street lights with colour temperatures below 3000K. The International Dark-Sky Association has also been vocal in this discussion, calling for street lights to be kept below 3000K and campaigning for them to be shielded, dimmed, or turned off at night when possible.

In response to this, and complaints from residents, many new conversions aimed for the warmer lights. The City of Montreal, for example, initially planned to instal 4000K street lights but then replaced many in residential areas with 3000K lights.

But elsewhere around the world, the first batch of LED street lights often remain in place. And new lights way above 3000K are still being installed. Changes in Rome may make international headlines, but who cares about the concerns of people in smaller, less-renowned towns? Falkirk at night is now lit by garish bright light – I don't know what the K is, I just know it's unpleasant. Likewise the riverside footpath in Lancaster is a no-go zone of extortionately bright light at 6000K. It's expensive to change street lights. If LEDs last anything like as long as they are purported to do, we may be stuck with them for a very long time.

The discussion and debate about LED street lighting is invariably focused on colour temperature, but Simon Nicholas says that lowering this will not solve all the issues and health problems, as the key issues of intensity and distribution of light are not being addressed. "It isn't being tackled because few understand it. I'm sure we will eventually see strict limits on luminous intensity. When the world wakes up, these street lights will be banned."

It is not just street lights that converted too soon, too bright and too blue. It's been the same story for public buildings and public transport across the world. The protest has burgeoned beyond social media into the mainstream press. *The Washington Post* reported that the New

York metro's original lighting was "buttery and retro and flattering", but the makeover was like "walking into a future where no one looks their best" and akin to "sitting inside a Xerox machine".

In the UK, the *Financial Times* reported on the blue-rich lighting in the Channel Tunnel trains: "Think of an operating theatre lit for intricate surgery or the most garish superstore with all overhead lighting cranked up to full brightness. As I did my best to get comfortable in this cold, flat environment, I rummaged for my sunglasses in my tote, hoping I'd packed them for the journey . . . Who wants to be blasted by cold, ghoulish light first thing in the morning, let alone later in the evening?"

Journalist Tyler Brûlé continues: "My local branch of Waitrose has also gone for a high-intensity lighting refit, and I can no longer face going inside. What was once a warm, welcoming interior is now inhospitable and forbidding. The surge in poor-quality lighting is not only a public health crisis, it can bring about a bottom-line crisis as well."

For architect and lighting designer Karolina Zielinska-Dabkowska, artificial lighting has become a "public health hazard": "I am convinced that the cost of this transition far outweighs the benefits," she writes in the science journal *Nature*. She has issued a clarion call for CFLs to be withdrawn from sale, LEDs to be more tightly regulated, and for physicists, engineers, medical experts and biologists to develop "biologically benign" forms of energy-efficient lighting. Natural daylight should be the focus, she argues, with policymakers encouraging and building regulations rewarding its use, especially in workplaces and hospitals where people have to spend a lot of time indoors.

Daylight: the ultimate low-energy light source is curiously absent from so much policy, regulation and discussion about lighting. And from practice – the switch to new forms of lighting is purported to be about saving energy, but there is so little emphasis on switching artificial lighting off altogether and using daylight whenever possible. I'm reminded of this every day when I take my daughter to nursery. We have a system in place where the staff see me coming and switch off the strip lighting in the foyer, so that I can help her hang her coat and change her shoes. I'm grateful for the access, but the moment I'm gone they switch the lights back on again, regardless of whether there is bright sunshine or a gloomy sky.

I get in touch with Karolina to ask why we don't talk more about daylight in building design. She explains that lighting, and especially daylight, was lacking in her architectural training. There are some specialist practices of daylight engineers, but architectural lighting is predominantly about artificial lighting.

I tell her that when I was born, my dad was teaching in Aberystwyth University, a 1960s new-build, where the main lecture theatre was partially underground and without natural light, and the three seminar rooms were all windowless. It felt so wrong, he says, but that was the thinking at the time – it seemed modern and progressive to rely on science as much as possible and nature as little as possible.

"Yes," Karolina says, "in the 1960s and '70s there were even schools underground and without windows! They were trying to prove that artificial lighting was better and you don't need daylight. Nowadays, there are requirements for windows of a specific size, and a percentage of façade is required to be windows."

Nevertheless, we agree, there still isn't nearly enough emphasis on the use of and need for natural light, both for human health and for real energy savings. Karolina has presented a paper at the UN World Sustainability Forum, which is focussed on improving human health and well-being across the world. She says that the discussion and goals are about healthcare, disease, access to water, but there is no mention of light as a fundamental component of health.

"The built environment is increasingly cutting off daylight. It's increasingly urbanised and people are spending ever more time commuting and working indoors," she says. "If we talk about 'skyscrapers', you'd think of Manhattan, but that's nothing compared to the new megacities in China and elsewhere in Asia. These are enormous skyscrapers combining offices, institutions and home – everything in these tall towers, but only the very richest will be able to afford a penthouse. If there is a square between the buildings, you still don't get sunlight there.

"We are running without understanding in a very, very wrong direction! Light is not part of these conversations!"

•

Incandescent

"The history of art is a history of looking at light."
– James Turrell

I visit Yorkshire Sculpture Park to see James Turrell's *Deer Shelter Skyspace* exhibit. Turrell is an American artist who takes the relationship between art and light to another level by making light the actual medium of his works. He talks of "the thingness of light" and of knowing as a young man that he wanted to create work that was not about light but that actually was light. He is known as a "sculptor of light" and his works are immersive: you enter them to contemplate space and light.

I enter *Deer Shelter Skyspace* with some trepidation, my claustrophobia fighting my curiosity – it looks closed from the outside, all red brick walls and concrete. I step closer, and there is a way in, guided by incongruous but reassuring flower-shaped lights on the floor. They lead me to a square room that has something of a tomb about it: there's just a concrete bench all around the four high white walls. I'm unsettled but lie down on the smooth concrete bench. And then I see it: above me is a square of sky. A perfect thing, this slice of sky, framed by the flat white roof. It's far above me but I feel like I could touch it, this exhibit of sky. It's changing slowly, wispy clouds break up the white ever so slightly, soft blues come and go, then three geese fly through the frame, calling. Then there's just white again, inside and outside the frame, a stillness, and the light itself is the thing of beauty.

Beauty. I've come round to thinking that aesthetics isn't

trivial, and beauty isn't a luxury but something fundamental to our health and well-being. And beauty is all about light. Try this: put a CFL in the bedroom and look at naked skin beneath it – whether your own or someone else's. Under the fluorescent lighting, skin looks thin, tinged with blueish-grey, cold, veined and dull. Then switch it off and light a candle. Skin is transformed: soft and warm, glanced with gold and enticing. So which is real? Which is true? Just like the colour of a painting. Just like the streets in Rome or Lancaster. Light doesn't just illuminate our surroundings – it forms our surroundings. It creates the very world that we perceive.

7

Banning the Bulb

The first I ever heard of it, I was on the beach in North Berwick, watching gannets dropping out of the sky and spearing the blue waves. Beside me, my mum had her headscarf pulled tight against the cold east wind and a stern, troubled look on her face. She wasn't watching the gannets, or the rolling sea. "*They* are going to ban normal light bulbs," she said in a grave pronouncement. There was a silence. That ominous, anonymous "they". Mum does a good line as a prophet of doom. I remember back in the 1970s, she was telling people that stuff in their hairspray could wreck the planet's ozone layer. No way, said my school friend's mum, I just can't believe that this little thing on my dressing table could make a difference way up there in the sky!

"It's going to make some people's lives very difficult," Mum continued. This was years before my light troubles began, and I didn't take much notice. No way, I thought. How could they ban *light bulbs*? There were bigger problems in the world to worry about.

Looking back now, I feel a sense of incredulity. How did it come to this? How have we been asleep at the wheel while artificial light has changed around us? How did something as simple and fundamental as an incandescent light bulb come to be banned across the world? On what actual basis was the ban formulated and justified? Did those making the decisions really talk about light?

When I ask, as I often do, "*Why* were incandescent light bulbs banned?" most people are shocked by the question and give me a blast of their own incredulity in a harsh stare. "Surely everyone knows? They were inefficient, they wasted energy, they were banned for environmental reasons!" But what *were* those reasons?

Others say, "Oh, but they were never *banned* – they were merely phased out when energy efficiency regulation was introduced which they could not meet." But this is a ban by effect, and specifically created to target incandescent lighting.

Yet some respond with a sardonic shrug, or raise their eyebrows and snort gently through their nose. "C'mon, you must know by now? Incandescent light bulbs were not making enough money! Patent expired, cheap to make, cheap to buy, high on popularity but low on profit. So make the case for new lighting technologies as low-energy alternatives and, hey presto, business is booming again – easy!" This sort of argument is often seen when any patent expires. But *how* could you pull off a feat like that, and convince the world this was a good plan?

There are actually two questions contained within mine: why *incandescent* lighting, of all the things you could ban if this really was about carbon emissions and climate change;

and why a *ban,* of all the ways you could bring about change, especially in a culture that's generally reluctant to go about banning things?

So I pick up the thread of this story and follow it back to the European Commission in Brussels, and a piece of legislation called the Ecodesign Directive. On 8 December 2008, the EU Commissioner for Energy, Andris Piebalgs, announced that the EU would be progressively phasing out incandescent bulbs, from 2009 to 2012. By enforcing a switch to energy-saving bulbs, he said, EU citizens would save close to 40 terrawatt hours – roughly the electricity consumption of Romania, or 11 million European households – and this would lead to a reduction of about 15 million tonnes of CO_2 emissions per year.

There are many routes from a light bulb to the tonnage of CO_2 being released into the upper atmosphere, and I'm curious to know which ones were taken to arrive at these figures. Because light, as I'm learning, is mighty complex stuff – as is electricity – and there are numerous factors to be considered, figures selected and choices made when compiling such a calculation.

Listening to talk about light bulbs, you could be forgiven for concluding that incandescent bulbs are inherently climate-change inducing, somehow secreting carbon emissions directly into the air. They don't, and so their contribution to emissions depends first and foremost on how the electricity is generated, and how much CO_2 is emitted in that process. Then, as different countries vary greatly in the percentage of energy made by fossils fuels or renewables, by which method do you calculate an average, and is that average meaningful?

Next, do you measure the electricity used for the entire product's journey – from mining and manufacturing components, production, transport, disposal and recycling – or only the electricity used to make the light when the bulb is switched on? If the latter, which type of each bulb is being compared? For example, the cover (or "double envelope") around many CFLs, which the EU advises for safety, also reduces the efficiency. And does a decline in light quality over the lifetime of a bulb affect the calculations for efficiency?

It gets ever more complicated because the energy used by the bulb is not the same as the energy drawn from the grid. CFLs have what is known as a low power factor, and although it sounds like a contradiction in terms, this means that the electricity companies actually have to supply *more* energy for their use – and this can also cause disruptions in the electricity network.

And if all that isn't enough, human behaviour also needs to be factored in. Studies from the UK and Sweden have shown that it's actually behaviour that makes the biggest impact on energy consumption with domestic lighting. How long will the light be on? How easily can it be dimmed or switched on and off? How does our perception of it affect the amount it's used?

Tweak any one of these factors, and you end up with a very different number. So who condensed all these into just a few set figures to justify a blanket ban? And, in the light of all these variables, what actually is an energy-efficient light bulb, and what can truthfully be called "eco"?

Trying to understand the ban: "Diving into a sausage machine"

How does one go about digging deeper into the decision-making processes of the European Union? The EU website offers downloads of all the legislation, but I want to know about the discussions that preceded it. Where did this information come from? Who calculated the 15 million tonnes of CO_2, for example, and how, and who scrutinised the calculations? I'm looking for some equivalent of Hansard in the UK, a record of the parliamentary debate. Ah, I discover there was no debate as such. This decision to ban incandescent lighting was never debated in the EU parliament, nor in the UK parliament. Apparently, that's not how things work. Instead there's a system called "comitology", in which committees rule the world. The European Commission formulates legislation, which goes on a journey through numerous committees consisting of representatives from member states, NGOs and industry, and is then put to MEPs to approve and ratify.

Which brings us to the Ecodesign Directive, an EU framework that aims to lower the energy consumption of electric products. The idea is that reduced consumption must be worked into products from the design stage, thus reducing greenhouse gas emissions by the resulting reduction of energy demand when the products are used.

The original 2005 directive covered groups of electric products such as machine tools, fridges, freezers and air conditioning systems. It had nothing to do with light bulbs. The directive was already rolling, and in the process of being extended for a 2009 update, when German Environment Minister Sigmar Gabriel proposed using it as a legal basis to ban incandescent light bulbs. He quoted the 15 million tonnes of CO_2 emissions that the EU would save each year by converting to CFLs. This would contribute to one of the EU's stated aims for 2020: to reduce energy consumption by twenty per cent compared to projections. In 2007, Gabriel wrote to the EU Environment Commissioner Stavros Dimas asking for quick action to implement this, and the idea swiftly hopped, skipped and jumped through the requisite committees on its journey to becoming legislation.

"Be warned," writes the German newspaper *Die Zeit*, "reconstructing the ban on incandescent lamps sometimes resembles diving into a sausage machine." According to its report, there was concern – "a sticking point" – about whether the mercury content of CFLs meant they really had a better impact on the environment overall, especially as so few consumers disposed of them properly. But the so-called "mercury paradox" was cited: that coal-burning power stations emit mercury and thus lowering energy demand reduces overall mercury in the atmosphere. And that was the matter settled.

A regulatory committee of MEPs did have the opportunity to submit the decision to ban incandescent lighting to parliamentary debate, but voted overwhelmingly against doing so. Mostly it was members of the Conservative group that wanted to continue the discussion. One Liberal group MEP, Holger Krahmer from Germany, who voted for further debate, said that he felt it was not right for the EU to take such a far-reaching decision in this way. It reminded him too much of life in the former East Germany.

Justifying the ban (1): Incandescents "a cause for concern"

So why light bulbs? A bit further back along the story thread in 2006 is the creation of a publication called *Light's Labour's Lost*, by the International Energy Agency. It bills itself as a guide to "Policies for Energy-Efficient Lighting," but in effect it's a 600-page dossier, forming a case against incandescent lighting.

It points out that when the incandescent lamp was first commercialised, the main mode of transport was the horse, trains were powered by steam, and balloons were the only means of flight. Why then, it asks, do we continue to rely on the "horse" of the incandescent lamp instead of the internal combustion engine of the CFL? Is it because we are inherently attached to these older technologies or unaware of the merits of the alternatives? Deploying cost-effective and higher-efficiency alternatives would allow for stronger and cleaner economies "without sacrificing anything in our quality of life".

The powerful language and big numbers continue

throughout the sizable tome. Under the subheading "Beacons of hope", it describes how "without a palpable change in lighting quality" a market shift to CFLs from incandescent lamps would cut world lighting electricity demand by eighteen per cent. But this change won't happen by itself: "a number of barriers limit deployment of lighting technologies" (including "market imperfections") and so government action is required.

Incandescent lighting is "a cause for concern" it says, as its efficiency is inherently low, with an energy to light conversion of just five per cent. The remaining ninety-five per cent is delivered as heat. Much of the discussion is phrased in terms of "lumens per watt" – the amount of light emitted per watt of electricity. Incandescent bulbs typically emit only 12 lm/W, while fluorescent bulbs emit between 30 to 110 lm/W. And while incandescent lifespan is on average "just 1,000 hours – significantly shorter than that of other alternatives", CFLs have rated lifespans of 5,000 to 25,000 hours. Based on all this and more, *Light's Labour's Lost* makes recommendations for policy and legislation.

In 2009, the EU commissioned an in-depth study from Flemish consultancy VITO to establish which Ecodesign requirements could be set for lighting products. The report assesses technical, environmental and commercial aspects of lighting. It notes that fluorescent lighting is exempt from the international directive banning mercury and other toxic substances from the EU market. CFLs containing no more than 5mg per lamp are permitted. While data for the environmental impact of mercury production are not available, mercury itself is considered an environmental impact. A sample of five CFL bulbs were tested for this study, which ranged from 0.28mg to 6.3mg:

> It must be stated that sample #3 significantly exceeded the maximum allowed mercury content. This is probably caused by the cheap but inaccurate method of mercury filling (drip filling) that seems to be very common in most small far eastern production plants.

Ideally mercury is collected when the bulbs are recycled:

> Collected CFLi's at end of life are crushed in a closed installation and sieved. The mercury containing fraction is distilled at 600°C to separate the mercury. The pure, metallic mercury is used again by lamp industry.

However, the report acknowledges that recycling rates for CFLs are low and reliable information is not available:

> At this moment it will be assumed that eighty per cent of the used CFLi's is not collected and as a consequence eighty per cent of the mercury present in lamps is emitted during the end-of-life processing.

Justifying the ban (2): "Health, safety and the environment shall not be adversely affected"

A technical briefing, prepared for MEPs at the time of the ban, is an interesting example of the big issues of selection when you're dealing with words and numbers. It begins, for example, by stating that "lighting may represent up to a fifth of a household's electricity consumption". Note the "may" and the "up to". CFLs are the "most energy efficient"; incandescents are the "worst". In summary tables of advantages and disadvantages, the energy saving of CFL compared to incandescent lighting is stated as eighty per cent. CFLs are described as "environmentally friendly", rather than energy-efficient, and incandescent is an "energy-guzzler".

As the legislation was being formed, there was already awareness that CFLs could cause health problems for some people. The technical briefing does mention that they contain mercury and emit ultraviolet light "which can, under certain circumstances, have a negative impact on people suffering from diseases accompanied by light sensitivity". However, it reassures, these two factors do not present a risk to the general public. It doesn't elaborate on what happens to those who are light-sensitive and thus about to be excluded from being part of the general public.

The lighting manufacturers themselves baulked at the speed of the proposed phase-out, due to take place between 2009 and 2012. "Whoa," they said, in effect. "We're on it – low-energy lighting is on its way – but production is not quite ready yet to supply that level of demand. Can we phase it a bit steadier?" No, said the environmentalists, climate change is happening now and we need to cut greenhouse gases now. We cannot waste a moment in this fight. "Their proposals go neither far enough nor fast enough," said the Green's Caroline Lucas, who called for a complete ban by 2010. "The reality is we don't need to 'improve efficiency' of incandescent lighting – we need to ban it as soon as possible," she said.

It seems the incandescent ban was built, and held high, upon four main pillars: a light to heat ratio of five per cent to ninety-five per cent, which was perceived as wasteful; an efficiency comparison with CFLs of twenty per cent to eighty per cent; a calculation of energy that would be saved by their absence and a consequent calculation of 15 million tonnes of CO_2 emissions from that energy; and a conclusion that despite the mercury content of their replacement, overall mercury in the environment would be lower without them. It was sweetened with the promise of financial savings for householders: that the longevity of the new lamps, and the reduced energy consumption, would outweigh the initial higher purchase costs.

And so it came to pass, in Regulation 244/2009 on the 18 March 2009, that incandescent lighting was legislated away from the twenty-seven member states of the European Union. Frosted or pearl bulbs were targeted first from September 2009. The phase-out was to be a rolling programme, with the most powerful 100 watt bulbs to go first in 2009, then 60 watt clear bulbs in 2011, and 40 and 25 watt by 2012.

The legislation contained some noteworthy clauses and qualifiers. For example, it states that "sustainable development also requires proper consideration of the health, social and economic impact of the measures envisaged". And that implementing measures shall meet all the following criteria:

— there shall be no significant negative impact on the functionality of the product, from the perspective of the user;

— health, safety and the environment shall not be adversely affected;

— there shall be no significant negative impact on consumers in particular as regards the affordability and the life cycle cost of the product;

— there shall be no significant negative impact on industry's competitiveness;

— in principle, the setting of an ecodesign requirement shall not have the consequence of imposing proprietary technology on manufacturers; and

— no excessive administrative burden shall be imposed on manufacturers.

Hang on a minute, did nobody say? *All* the following criteria. Does the change from incandescent lighting to CFL actually meet *any* of these criteria?

Fighting the ban: "This mad musical-chairs of gesture politics"

It is not that the ban wasn't noticed – or indeed contested – but this was not enough to halt, or even pause, the legislative process. The very concept of banning incandescent lighting had become some kind of juggernaut that was already rolling around the world by its own momentum. Most protests happened after the swiftly enacted ban was in place. But even before the ban and as the phase-out began to bite, there were ripples of protest throughout Europe, in parliaments, in the press, among medics, and among artists. Concerns were raised about democratic process, ideology, human health and environmental impact.

"Dimwits!" railed the *Daily Mail*. "There was not a hint of democracy in this crackpot decision. Have these politicians got the faintest idea what they are talking about? Do they actually look at the hard, practical facts before they rush to compete with each other in this mad musical-chairs of gesture politics?"

"Eco-light bulbs aren't what they seem," said BBC Radio 4's *More or Less* in an analysis of the comparison between incandescent and CFL on energy efficiency, brightness and longevity. It quotes European Commission advice that "exaggerated claims are made on the packaging about the light output of CFLs" and points out that brightness is tested in laboratory conditions, but within the home fluctuation in temperature can reduce CFL brightness. Further studies show that CFLs can get twenty per cent dimmer over time. The lifespan is based on an average – some may last the promised ten years, others won't. This figure is calculated by assuming a light is on for three hours a day, but frequently turning the light on and off can more than halve the expected lifespan. Saving eighty per cent more energy is also an exaggeration: it's an "up to", depending on which bulbs are used.

The Spectrum Alliance was founded in the UK to campaign for an exemption to the EU lighting ban on health grounds. This was an umbrella group of medical charities whose members reported that their conditions worsened by the new fluorescent lighting. These charities included Migraine Action, the National Autistic Society, Lupus UK, Electrosensitivity UK, the ME Society and the XP Support Group. The Spectrum Alliance calculated that more than two million people would be adversely affected by the ban, and met with politicians and civil servants in the UK and EU to argue that light-sensitive people must have access to incandescent light bulbs.

"There are clearly wider implications of the ban in that low-energy lighting makes workplaces and public places inaccessible to those affected. However, it is a fundamental human right that our homes are safe for us. If we cannot buy incandescent light bulbs then this is not the case," wrote Spectrum Alliance Coordinator Catherine Hessett in a letter to the Equality and Human Rights Commission. Catherine campaigned widely and raised the profile of the issue in parliament. She received assurances from ministers that the government would "work with stakeholders to avoid any unintended adverse impacts" and that "EU legislation is unlikely to ban all incandescent lighting products except where there are suitable alternatives." All politicians involved in the ban were well briefed on its effects, Catherine said.

One passive form of protest was to hoard. While you could still do such a thing as pop down to the hardware shop and pick up an incandescent light bulb, people did so – stocking up for the years ahead, sometimes with a lifetime's supply. The German press reported widespread *hamsterkäufe*: panic buying and stashing of incandescent light bulbs. I quite like the idea of being a *glühbirnenhamsterin*. I began to keep a list of reports of fellow hoarders. Dermatologist John Hawk. Ophthalmologist John Marshall. Dance lighting designer Michael Hulls. German pop star Nena. US lighting designer Howard Brandston. My mum. Benny from Abba, so I've heard. Bob Dylan, perhaps? He famously responded to a journalist's question about his true message: "Keep a good head and always carry a light bulb." Apparently he lists an incandescent light bulb in the dressing room as a requisite on his backstage rider.

Two enterprising Germans set up a company as a form of what they called "action art". It played on the much-quoted accusation of inefficiency in incandescent bulbs – that they used only five per cent of their energy as light, and "wasted" the rest as heat. Ah, but there are no restrictions whatsoever on using energy for heat, argued one of the partners, engineer Siegfried Rotthaeuser. Under the brand name Heatballs, they imported incandescent light bulbs as "small heating devices", which happened to have a useful by-product of making five per cent light. It was always meant as a protest, they said, a political satire as a "resistance to the legislation which is implemented without recourse to democratic and parliamentary process". A percentage of the sales would be donated to rainforest conservation, which seemed a better way to save the environment than "toxic CFLs". The Heatballs initially sold well, but the imports were soon impounded by German customs officials. According to Rotthaeuser, "the question of why an electric patio heater is allowed in bar form and a bulb-shaped heater is prohibited remains unanswered".

In Germany, an influential group of artists published a statement in the *Neue Westfälische* newspaper, saying that incandescent light is "indispensable" and demanding that the decision be revoked: "We artists, exhibition designers, museum curators, architects and designers cannot accept the EU phasing out the incandescent lamp." The paper goes on to point out that the consumption of energy in homes has risen since the ban, and that the savings effect of the ban is "zero", quoting the UK Department for Environment, Food and Rural Affairs: "When the heat dissipated by conventional incandescent lamps is reduced, the missing warmth is compensated for by the central heating system."

With the 2011 documentary *Bulb Fiction,* Austrian director Christoph Mayr explores many paths in the light bulb labyrinth. It shows CFL-related mercury poisoning in a five-year-old boy in Germany who had broken a bulb in his bedroom, and in a trainee in a recycling plant worker in England. Workers in England and India are featured recycling CFLs without protection, unaware that they are handling a dangerous substance.

In the film, Ravi Agarwal of the charity Toxics Link describes the situation as being about the desire to feel good without dealing with the "deep and difficult real issues" of climate change. He says that what people don't realise about mercury is its dual state as a liquid and a gas – they think if they can see it on the table it isn't also in the air. Mercury from CFLs has seeped into the water table and pollutes the wider environment, he says, and it's the same story in Africa, in Southeast Asia, and Latin America . . . everywhere.

Christoph Mayr also visits the archives in Berlin, which holds thousands of documents from the Phoebus Cartel of the 1920s. Lighting manufacturers colluded to ensure that incandescent bulbs didn't last longer than 1,000 hours, with fines and punishments for any company who extended the lifespan. In the 1980s, a German engineer filed a patent for incandescent bulbs lasting 150,000 hours, and his invention was trialled in Berlin streetlamps, but he died suddenly before production could be established.

Questioning the new technology: "The basis for the ban is illogical"

Lighting designer Kevan Shaw was among the first to sound a warning to fellow professionals about losing a basic tool of their trade. Always with an ear to the ground and an eye on future developments, he saw it coming, and foresaw the impact. For him, "the basis for the ban is illogical. The colder the country, the more non-fossil fuel electricity that exists, the poorer the environmental argument is for swapping incandescent lamps for CFLs. The industry are . . . removing a good, simple, practical and safe product that is cheap to produce and cheap to sell [and replacing it] with a variety of expensive, complex and more polluting things."

Interviewed in *Bulb Fiction*, Kevan explains that the so-called "mercury paradox" doesn't hold up when you subject it to closer scrutiny, the numbers involved having been spun to prove that CFLs are not a problem. The figures used are based on US coal, which generally has a higher mercury content than coal in Europe, and they rely on CFLs lasting a certain number of hours (which they rarely do in practice). But the real issue is what happens to mercury when you burn coal – in a power station in Scotland the mercury is well contained, whereas most mercury from CFLs ends up in landfill.

Kevan has spoken, written and campaigned extensively on this issue, and has now been embroiled in it and the legislative process for more than a decade. I want to know what motivates him.

"It's wrong," he says.

Wrong? Like, wrong factually? Aesthetically? Morally?

"All of them. I also feel a professional duty. Because of my work I do understand light, and I feel I need to inform others who maybe don't."

Kevan talks about a "huge lack of understanding" about light, which is as much the case with legislators as it is with the general public. He describes consultation meetings in Brussels at which the civil servants are completely out of their depth on the fundamentals of the technology being discussed. And he points out that when renewing or updating the regulation, there has been no discussion or apparent need to prove that existing regulation has been effective in its core purpose of saving energy. Surely it's vital work, he argues, to find out if there are actually energy savings involved?

"And reducing energy use and reducing CO_2 emissions are not the same thing – it depends on the generating capacity used in peak lighting hours. In many EU countries this remains coal-fired, so surely what's needed is the regulation of energy generation not lamps!"

Nor is "lumens per watt" a meaningful way to measure light, let alone legislate about it. "It's a very, very basic crude measure," he says. "We don't *see* in a quantity of lumens or lux. The concept of lumens came out of candle power and measurements of gas light in comparison with a candle. This whole method of measuring light in terms of the light source has been entrenched for a very long time. We need to chuck all that out and start again, looking at the different requirements and functions of the light and basing it on what we actually see."

Kevan also explains how European manufacturing suffered from the speed that the legislation was enacted: turbo-charged by environmentalists, the ban came into force so fast that the European lamp industry could not supply the demand. Factories in China and the Far East could, however.

"Because the Chinese costs of manufacture were lower – lower labour costs, lower health and safety around workers, and at that stage the Chinese were still building coal-fired power plants to provide relatively low energy costs – there was no way EU manufacturers could make CFLs, and they very quickly started closing lamp plants."

What do you make of it all? I ask Kevan. *If you step back and look at the whole story, what's the main conclusion you'd draw from it?*

"That it's a drop on the ocean," he says.

How do you mean?

"*We* are learning what is happening to light because I'm working with light and you're affected by it. But the same story is happening to everything – we probably don't see it, just like others don't see light."

Do you sometimes feel like the wee boy in The Emperor's New Clothes?

He looks nonplussed for a moment then smiles a wry smile. "Yes! All the time. Exactly that!"

So, have I got this right? I ask. *The legislation didn't comply with its own criteria, the green movement campaigned for a potent environmental toxin, and it was all supposed to be about saving energy, which there is no evidence that it has . . . ?*

"You should write a satire," he says bleakly. "It's the only way to tell this story. Go through it all, step by step, and then say at the end: 'This actually happened.'"

In Sweden, Inger Glimmero Nordangård's professional world also led her to notice, question and contest what was happening to light. She was working in Stockholm as a colour consultant in interior design when the incandescent ban rolled in, and she was appalled by the rendition of colours under CFLs.

The ban on incandescent lighting is wrong on so many levels, she says, as she feels it's based on information that is completely flawed: "It all depends on twisting numbers any which way to make a chain of faulty calculations. It's all assumption – like that the electricity comes from coal. Is there any argument against an incandescent bulb run on solar panels? But once these assumptions are put out there by environmental organisations, it turns into fact. Then the ideas spread through organisations and propagate themselves. Nobody checks! Only nerds like me go through the figures – most people don't have the time and energy and so just presume it must be true.

"It's clear that the incandescent phase-out is invalid by its own criteria!" she continues, running through the Ecodesign Directive:

"No significant impact on the functionality? With CFLs and LEDs, the user gets poorer light quality, suboptimal colour rendering, often dimmer incompatibilities, to name but a few issues.

"Health, safety and the environment shall not be adversely affected? CFLs contain mercury, emit UV radiation. Many LEDs emit harmful amounts of blue light. The EU itself acknowledges that the health of 250,000 people will be affected.

"No negative impact on consumers regarding affordability and life cycle costs? CFLs have been hugely subsidised by tax payers' money – our money! With CFLs and LEDs, dimmable, higher quality lamps are expensive to buy and claims of longevity are exaggerated. Both become dimmer over time. And has anyone ever calculated the true life cycle cost? The mining of mercury, phosphors and rare

earth metals, the transport around the world of electronic and chemical parts, recycling of toxic waste, depositing mercury – which is now banned from EU export."

We talk of all electric light as being "artificial light", but Inger describes both CFLs and LEDs as even more artificial: "It's *synthetic* light. Incandescent light is *real* – it's contained fire. Banning incandescent light is the equivalent of banning water in order to force everyone, including diabetics, to drink only Coca-Cola when they are thirsty. Light is a bio-nutrient, just like food, air and water, and good light quality should be a basic human right."

Reviewing the ban: "It's another inescapable loop"

What can you do, should you happen to be a light-sensitive person and EU legislation takes away a product that you need and offers no provision for you to be able to light your home and live your life without pain and ill health? People wrote to their MEPs, who can communicate on behalf of constituents by submitting questions. But the replies kept coming back asserting that there was "enough choice available" for light-sensitive people. You get caught in a loop – however you phrase the question, the same reference is used for the answer, and there seems to be no way to challenge that.

All answers on health referred back to the findings of the EU's Scientific Committee on Emerging and Newly Identified Health Risks (see Chapter Four). In writing to MEPs, many correspondents received the response that

"SCENHIR concluded that suitable alternatives to incandescent light bulbs already exist for all light-sensitive people". This has always been contested in practice, as some light-sensitive people find that incandescent light is the only artificial light in which they feel well. But then the alternatives, and consumer choice, were further constricted in a series of updates and reviews of the original legislation.

Some incandescent light bulbs had continued to be available through a loophole to allow "specialist lamps". These included "rough service lamps" purportedly for industrial purposes that required incandescent lighting (which could withstand heat and vibration), and decorative "squirrel cage" lamps, which quickly spread throughout fashionable pubs and restaurants. Legislators seemed irked by these supposedly specialist lamps becoming "mis-sold" for domestic use, and so closed the loophole through an amendment in 2015. Only incandescent lights for ovens and traffic lights remained as specialist exemptions.

There was no remit for SCENHIR, or any other committee, to conduct, initiate, mandate or fund any further research. Its role was to survey existing scientific papers. And so if the research on a particular aspect of light and health had not been done, the conclusion would be that there was no evidence of a problem. And if there is no evidence of a problem, there is no impetus to conduct or fund further research. It's another inescapable loop in which light-sensitive people become trapped and invisible.

What wasn't reviewed, three years in from the original

legislation, was whether it had achieved the promised goals of extensive energy saving. In 2012, David Martin MEP placed a written question to the European Parliament, quoting the promised savings of 15 million tonnes of CO_2 emissions each year and asking what monitoring had taken place to measure the effectiveness of the regulation and for proof that this was being achieved. The reply came that it was "still premature to draw conclusions".

Globalising the ban: "Prison for possession in Zimbabwe"

I first discovered the truly global extent of the ban while chatting online. There's an ongoing theme – part jest, part desperation – of the need to escape the constant encroachment of new lighting, and we scan the world for a place to build a refuge. I spot something in the paper about uninhabited Scottish islands for sale, and post it, suggesting that we set up an incandescent commune there.

— Ah, but wouldn't that be subject to EU import and manufacturing law?
— Hmm, good point, let's get outta Europe altogether, somewhere far away. Australia?
— No, Australia was one of the first to bring in a ban! They kicked this whole thing off!
— Oh. New Zealand, too?
— More hope there. Kiwis were all set for a ban under a Labour government in 2007. Then the next

government shelved it in 2008. For political reasons, not because CFLs were shite lights.

— To piss off their predecessors?

— Or to piss of the Aussies!

— Okay, somewhere in South America?

— Hmmm. Argentina banned sale and import in 2011, after handing out 25 million free CFLs in exchange for incandescent bulbs, which were destroyed . . .

— Brazil started a phase-out in 2012, then total ban in 2016 . . .

— Cuba?

— No, they were the first in the world! The film *Bulb Fiction* shows enforcement officers going to houses, smashing incandescents and replacing them with CFLs.

— Are we going to keep this going through the alphabet . . .?

— There is no "D" in South America.

— Denmark!

— Er, that's in the EU.

— Yeah but what about all that *Hygge* stuff, soft lighting and cosiness that's supposed to be an integral part of being a Dane. How can they cope with CFLs in the home, or harsh LED street lights?

— Ha, dunno.

— Switzerland? They're not in the EU.

— Yes, but they're bound by trade deal to the European ban.

— Zambia?

— Nope. There are exchange programmes all over Africa to change from incandescents – millions invested to encourage the phase-out.

— Just stay clear of Zimbabwe! You can be fined for *possession* of incandescent bulbs.

— You are kidding?

— Check it out: ban wasn't working well enough, there's now a two-year prison sentence for possession. It's like they're some kind of dangerous drug. Again, incandescent bulbs were confiscated and destroyed on the spot . . .

— No!

The banter is light, but the fear is a leaden weight. A world without incandescent lighting is, for some, a world with nowhere to go after dark, no prospect of living a normal life. And again, I feel that niggling sense I had at the very beginning of this story that something here doesn't sit right. Why this global demonisation of incandescent light bulbs? Why the evangelical promotion of CFLs in countries ill equipped to deal with mercury in waste? Why does it cost millions to promote a new form of light bulb? Who gets that money? Why such absurdly draconian measures in Zimbabwe? What posturing is this, and to what or to whom?

I'm startled by a headline on an environmental website: "UN wants to help countries banish incandescent bulbs once and for all". I scan news reports of the ban throughout the world and find constant references to three bodies: UNEP, GEF and en.lighten.

As a result of a US$25 million UNEP project in Vietnam, "more than 60 million incandescent light bulbs have been phased out and replaced by domestically produced CFLs".

"Supported by a GEF grant totalling US$1.82 million this project will help disseminate 400,000 CFLs" for the city of Lome, Republic of Togo.

"Some fifty-five countries have joined the en.lighten initiative", including Chile, where "a transition away from inefficient incandescent lamps . . . would save 1.2 million tonnes of CO_2".

UNEP is the United Nations Environment Programme, which sets the global environment agenda. GEF is the Global Environment Facility, founded initially as part of the World Bank but now an independent organisation. (One of its six stated focus areas is toxic waste, especially persistent organic pollutants – POPs – and mercury, which accumulate and enter the food chain). En.lighten was a multi-million dollar UNEP programme, which ran from 2009 to 2017, also known as the "Lighting Efficiency Accelerator", with the objective of globally phasing out incandescent lighting by 2016. Funded by GEF, it was described as a public-private partnership between UNEP and the lighting manufacturers OSRAM and Philips Lighting. Sixty developing countries joined, which, according to its website, would save 35 million tonnes of CO_2 annually.

En.lighten ended in 2017 as a separate initiative and became part of United4efficiency (U4E), another public-private partnership under the auspices of UNEP, with a broadened scope from lighting to other technologies. Its manufacturing partners include OSRAM, Megaman and Signify (the new company name of Philips Lighting). It describes itself as "a global initiative led by UN Environment, funded by the Global Environment Facility and supported by companies and organisations with a shared interest in transforming markets . . ." It is now urging developing countries to "leap frog" to LEDs, for higher energy savings and to avoid the use of mercury in CFLs.

Ban in the USA: "This violates the very principles upon which this nation was founded"

And the USA. Even the USA. How can a country where consumer choice is so utterly sacrosanct come to ban light bulbs? But it did. The 2007 Energy Independence and Security Act (EISA) brought in a rolling ban of "inefficient light sources" from January 2012 onwards.

It seems there was more resistance than in Europe – mostly ideological protest at the violation of freedom of choice, some patriotic pride in the achievements of Thomas Edison, but also some practical concerns about CFLs as a fire risk. There are more widespread concerns about the fire safety of CFLs, as the electronic ballast in the base still tries to function at the end of the lamp's life, potentially leading to overheating. But this has been more of a problem in the US, whether because there is twice the current going into the lamps, or that there are more recessed and enclosed fittings that create ripe conditions for overheating. The US Consumer Product Safety

Commission has recalled specific brands that have been involved in house fires.

Leading the ideological dissent were Texas senators, who first tried to overturn the legislation and then brought in a Texas House Bill in June 2011, which repealed the lamp ban to allow manufacture and sale within the state. Indeed this was encouraged as a business opportunity, with senators wooing lamp manufacturers to set up a factory in Texas. This provoked both angry and envious responses from other states. "Texas tells Feds: Shove your light bulb ban," reads a 2011 headline in the conservative *Illinois Review*.

In 2011, lighting designer Howard Brandston was called to testify to the Senate Energy Committee about why he felt banning incandescent lighting was inappropriate. In his testimony he appealed to patriotic sentiment: "I believe this violates the very principles upon which this nation was founded and I, as a devoted citizen, am most proud of, our freedom of choice in our personal lives."

He continued: "I firmly believe that the restrictions put on incandescent lamps will have a significant negative impact on almost every residence in our country." He also expressed health concerns about CFL lighting, especially for people with auto-immune conditions such as lupus, and about their mercury content: "This 2007 light bulb standard brings a deadly poison into every residence in our nation." He asked the committee to consider more conceptual issues: "Lighting is not a product – it is a system designed for a purpose. This act separates one component of that system, the light source, and that destroys the success of the final design."

Howard tells me that he followed all the Senate rules: "I submitted testimony in writing a month in advance . . . and I read my testimony into the record verbatim. But during the question and answer session afterwards, there are no rules and you can say whatever you want. I went first and I got up and said, 'I'm very pleased and proud to say I'm the only person who paid my own way to get here and who does not represent an interest group.' . . . I got an interesting debate going. It's very hard to debate with me on light and lighting, I'm a real pioneer and have been doing this for sixty-six years!"

I noticed when I contacted him that Howard's email begins with "LightPain@", and I wonder, given his articulate outbursts about CFLs, if he has suffered pain or health problems from them himself?

"No, ha!" he chuckles. "I've been banging my head against the wall for years, I was such a pain in the side of the lighting manufacturers that they called me 'light pain' and I made it my email address! When the Department of Energy people see me coming they walk to the other side of the room . . . But no, no ill effects personally. I just don't like what lighting with CFL does to the places it's lighting. A lot of people don't see it that way. I'm very sensitive to the quality of illuminated space, and that quality is reduced.

"But I am concerned. We grew up with daylight, with the light of the sun and a total spectral power distribution – we grew up *seeing* like that. The eye has not been trained to see the way these new lights light things up. And I'm concerned in another decade or so when all these burnt out LEDs and CFLs go into landfills they will pollute the water. These light sources should not be tossed in the trash! They should be recycled, but people don't. In landfill the mercury and other toxins will seep into the soil and water system, poisoning the people. I probably won't be around to see that as I'm getting on in years . . . but this is still a crusade of mine. I will not let go!"

Moving the ban forward: "How can it be claimed that the general population does not include the old, the young and people with health problems?"

Meanwhile, back in Europe, the legislative wheels are now rolling into a new phase of the lighting ban. As I write, the EU is in the process of streamlining the different aspects of the Ecodesign Directive that apply to lighting into a new Single Lighting Regulation (SLR). This will tighten specifications for lighting still further and ban almost all remaining incandescent light sources – although it's unclear whether, or to what extent, Brexit will affect the implementation of the SLR in the UK. The SLR will also ban most halogens and many fluorescents – including CFLs themselves, in line with more stringent restrictions on mercury. The practical effect of all this will be to leave LEDs as the dominant form of lighting available for inside and outside spaces.

Early drafts of this legislation removed many of the

categories of special purpose lamps, including previous provisions for people with photosensitivities and for theatrical lighting. This sparked new protests, as did continued concerns about the safety of LEDs. The charity LightAware was founded in 2015 and took over from the Spectrum Alliance in the role of advocating for a new exemption to the lighting ban, which would allow light-sensitive people to access the incandescent bulbs they need. LightAware points out that the EU's own report by the Scientific Committee on Health, Environmental and Emerging Risks (SCHEER) had concluded that although the blue light emitted by many LEDs was generally safe, about a third of the population – the over sixty-fives and the under-eighteens – should be considered "vulnerable and susceptible populations" with regard to blue light. LightAware has raised concerns that the new legislation was ignoring these warnings in the rush to convert the continent to LED.

"We consider the SCHEER report's conclusion that the blue-rich LEDs currently in common use are 'safe for the general population' to be intellectually dishonest," says LightAware trustee Dr John Lincoln. "However, the report fully acknowledges that there is a problem with blue-rich light for older people, younger people and people with certain health conditions. We ask, under what circumstances can it be claimed that the general population does not include the old, the young and people with health problems?"

I discover too late that there was a public consultation on the new legislation. There is a stage in these procedures when theoretically any member of the public can download the draft documents and comment on them. But, of course, most people simply don't know about it.

What does hit the headlines is the Save Stage Lighting campaign, run by lighting designers who galvanised again, with extraordinary passion and energy, to call for an exemption in this legislation for stage lighting. "Save Stage Lighting" was projected onto the outside of theatres and concert halls across Europe, and a petition gathered more than 80,000 signatures. What caused such a reaction was the realisation that the legislation as proposed could literally leave theatres and other venues in the dark. For some lighting needs, the technology doesn't even exist yet to meet the stringent regulations. And the law would be financially devastating to all but the largest venues as it would involve replacing entire lighting rigs. The campaign also pointed out the absurd waste of such an immediate ban rendering perfectly good equipment obsolete.

Together the arts world plunged into the fray with a powerful and highly articulate case for the continued use of their lighting tools, including incandescent lighting. The campaign secured an exemption on a number of products for stage lighting. But the story was presented in the media as if stage lighting was the only crisis induced by this legislation. Somehow it was forgotten that all the world's a stage, and the scenes beyond the theatre deserve the same.

Lighting designers don't use the word "lights" in the way the rest of us do, both for the actual bulb and the contraptions they fit into. They talk about "lamps", "fittings" and "luminaires". This is more significant than it sounds, for one element of this global story is a semantic confusion between an energy-consuming electronic product and the stuff that comes out of it. We talk of "a light" as an electronic household appliance, and, as such, light bulbs are governed by laws about products such as hoovers, washing machines, kettles and irons. If a low-energy kettle boils less water, it doesn't change the nature of the water. And using it is an individual choice. But the light that comes out of a low-energy light bulb is not the same as that from an incandescent. And light is not personalised – it is a shared part of an experienced environment. The output from a light bulb is not a product, but a force of physics, affecting our bodies at a cellular level and our souls at an even deeper level. And yet it is regulated as a product: in the UK, lighting regulation is the responsibility of the Department of Business, Energy and Industrial Strategy – not the Department of Health.

I don't understand how it has been accepted on a global scale that a minimal measure of energy efficiency is the only criterion by which decisions are made about artificial light, ignoring quality, aesthetics, health and total environmental impact. Is this thinking applied to any other aspect of life? In food, transport, leisure, building and communications, wider criteria are considered.

You can download a cumbersome mountain of documents, and follow hundreds of pages of legislation, but none of it answers the unanswered questions:

— How can it not matter that legislation doesn't meet its own criteria?

— How can legislating bodies continue to make huge changes to artificial lighting, while acknowledging fundamental gaps in our knowledge about light and health?

— How can legislation continue to tighten without any evidence of the efficacy of its original intent?

— How will people who cannot tolerate LED lighting be able to live their lives?

— How does LED lighting affect the health of all humans and the environment?

— Last but not least – sorry to have to ask again, and to keep on asking – but I still haven't heard a satisfactory explanation: why was incandescent lighting banned?

I'll leave the last word in this chapter to Minna Gillberg, former advisor to the EU Environment Commissioner Margot Wallström: "The environmental benefit we are talking about here is really a political advantage. From this perspective, the environmental benefits we speak of become an expression of a symbolic climate policy that benefits neither the environment nor human health."

8

The Language of Light and an Ideological Tangle

"It's like you've got dog shit on your shoe," says Justin.

Sorry? I'd clocked the simile but still couldn't help myself twisting my foot for a surreptitious glance at the sole of my trainers. *Go on, I'm listening,* I say. *I need to know how this looks from where you're standing.*

We're sprawled in the sun on an early summer's day in Edinburgh's Princes Street Gardens, pink blossom scattered all over the bright green lawns. I'm pissed off with him but I can't help smiling. Justin's a visionary kind of guy – brave, sussed and astute – one of those who sees a bit further than the rest of us, and thinks a lot deeper. He can see the links and layers between things and works tirelessly for social and environmental justice. Yet I'd detected a sneer when he'd heard about my "light bulbs stuff". He's not the only one, of course – I'm still aware that hackles and eyebrows rise when I bring up the issue – but I reckoned he'd be the most able to articulate where that's coming from. And so I pinned him down and asked him straight.

I've got a story here, I say, *about injustice, equality and the environment, and I get a sense that you – like many of my friends – can't hear me, just don't want to know. It's like a barrier comes up and I'm on the wrong side of it. I need to stop getting upset about that and understand where it's coming from. Explain it to me.*

He nods, thinking, and squirms a bit, beginning to grasp it. And that's when he explains why it's like I've got dog shit on my shoe. He's only heard of complaints about the incandescent ban coming from the right, and the far right, and climate change deniers, and so that kind of thinking has somehow "owned" opposition to the ban.

So, in opposing it too . . . what I have to say is kind of repulsive by association?

"Yes, exactly that."

Hmmm. I get what he's saying, but still . . . *I oppose the ban on incandescent lighting. I think it's wrong. Donald Trump has expressed opposition to the ban on incandescent lighting. He thinks it's wrong. Does that somehow equate me with any aspect or ideology of Trump? C'mon.*

We agree that it's an inconvenient truth – another one. And we note the irony that in raising awareness about the problems of low-energy lighting, I share with climate change campaigners the enormous challenge of trying to tell people something that they just don't want to know.

I understand what Justin's saying; that the anti-ban sentiment seems to belong to an ideology far from his. And indeed mine. When I began to rummage around the internet and to scratch below the surface of this story, I wasn't prepared for the level of vitriol in web discussions, particularly in the US, where climate change deniers ranted about a conspiracy between environmentalists and "Big Government". Here's a flavour from one site:

— I just spent an hour in Lowes looking for "real light bulbs" so that I can see again. Had no idea the government had taken this source of happiness away . . .
— I have an idea. Keep using them and tell the gov to shove the curly cue bulbs up their ass.
— Climate change is a lie. Don't let the environmental Nazis destroy your quality of life. These evil bastards want to go after your car, your clothes, even the food you eat. Some eco Nazis want to make eating meat illegal or expensive because cows generate a lot of carbon. Environmentalists are the enemy, they are anti-capitalist, they see humans as parasites, as entities that need to be removed.

And another:

— Save money: let fat, black, angry Commie-Libs invade your home and throw away brand-new incandescents.

Follow the comments below media articles on the issue and you'll find much the same tone. Is this what led to the assumption I encounter here in the UK, that those opposing the ban don't believe that climate change is real and critical?

From another point on the ideological spectrum, legislating on light bulbs is an affront to the free market ideal: the US government has done something fundamentally anti-capitalist in using legislation to interfere with market forces. In his polemic *I, Light Bulb*, Michael Patrick Leahy of the Nationwide Tea Party Coalition pleads the case of an incandescent light bulb on death row "sentenced to die by the hand of Big government". He complains that government has colluded with industry, and he argues that all light bulbs should have been allowed to "duke it out in the marketplace", where the incandescent bulb would prosper against the "depressing, color-draining, sickly, headache-inducing" light of its opponent, the CFL.

I read plenty of criticism of the ban coming from centre right publications, which also make the case for capitalism. And yet there is also anti-ban sentiment from anti-capitalists, arguing that this is "greenwash", a wolf in sheep's clothing – the insidious corporate takeover of light masquerading as environmentalism.

In the UK media, the commentary around the ban is often ideologically loaded. Take this *Telegraph* headline: "Official responsible for light bulb ban is a former communist". Or this *Express* piece: "What the light bulb ban has done is to treat consumers like idiots. Like all EU laws it was done behind our backs, by commissioners influenced by vested interests and without proper public debate." The most vociferous protests and questions have come from the *Daily Mail*, which gave out 25,000 incandescent light bulbs in 2009 as a protest against European interference. It appealed to British patriotism and nostalgia: "Throughout war, disaster and recession, it has kept Britain illuminated for more than 120 years. But the traditional incandescent light bulb is finally being switched off for the last time . . ."

The *Guardian* went to town in the other direction: "Good riddance to incandescent light bulbs," said the science correspondent in 2009, dismissing any resistance as "an outpouring of ill-informed nostalgia" from "backward rose-tinted doomsayers" and "stick-in-the-muds". A further 2009 article said that the incandescent bulb "will join tape cassettes and typewriters as defunct consumer items" to make way for "environmentally friendly" CFLs, a change which is a "no-brainer", and those opposing it need to be weaned from their "planet-spoiling behaviour". Columnist George Monbiot lumped incandescent light bulbs with patio heaters and garden floodlights as "wasteful and unnecessary technology" and called for their immediate ban.

Where health concerns fit the ideological critique of the ban, the coverage is extensive: it's the perfect ammunition. At the beginning of 2008, the *Daily Mail* was already warning "Evacuate the room now!" if a light bulb goes, elaborating on the mercury risks and the lack of labelling, before moving on to wider health concerns, including light-sensitive skin conditions. Over the years since, the *Daily Mail* has run a number of in-depth articles, with leading medics expressing their concerns about the health impact of new light bulbs: "Low energy lamps cause rashes and swelling" in 2009; "Energy-saving light bulbs can fry your skin" in 2012"; "Do environmentally friendly LED lights cause BLINDNESS?" in 2013. The *Telegraph* also flags up health concerns, highlighting that some people were stockpiling because the new lights caused migraines.

Incandescent

The *Guardian* dismissed health concerns as "scare-mongering" and a "notion": "There is no scientific evidence that they are caused by CFLs." If there are real problems, it's "clearly only a tiny minority" and so "not a reason to maintain the status quo". It has since been largely silent on health issues and the injustice of the social exclusion of light-sensitive people. It just doesn't *fit*; it's ideologically inconvenient to get ill from lighting that is believed to be Good and Eco.

•

Somewhere in the 1990s, the language of light began to change and the incandescent light bulb seemed to acquire a mandatory prefix. It was no longer sufficient to use the word "incandescent" without selecting from a choice of adjectives to precede it: old-fashioned, obsolete, wasteful, energy-guzzling, inefficient. Inefficient incandescent, inefficient incandescent – say it often enough and the words join together and weld somehow until nothing can prise them apart. Just when was it decided – and by whom – that one very narrow gauge of efficiency was the only measure by which we would assess light bulbs?

The incandescent light bulb became a scapegoat for the mess the world is in. Language was harnessed to accuse it, and its sacrifice seems to placate something. The dialogue was firmly set and it felt like two sides formed around it with a clear line between them. Alongside the environmental campaigners – the Green Party, Greenpeace, WWF and Friends of the Earth – were politicians such as

Gordon Brown ("Our aim is that every householder installs low energy light bulbs"), German Social Democrat Sigmar Gabriel ("Ambitious efficiency criteria for lights need to be introduced") and President Barack Obama ("I will immediately sign a law that begins to phase out incandescent light bulbs").

On the other side, we find UKIP politician Paul Nuttall: "This is yet another example of ill-thought-out legislation made by the EU before they have got all the facts." Or Conservative politician Martin Callanan: "Rather than forcing people to switch by removing their choice we should be using the right incentives to encourage change." In the US, there was The Tea Party Movement: "The American people want less government intrusion into their lives, not more, and that includes staying out of their personal light bulb choices," said Republican Michele Bachmann. And @RealDonaldTrump: "Remember, new 'environment friendly' light bulbs can cause cancer. Be careful – the idiots who came up with this stuff don't care."

The steady ground beneath my feet began to tilt and shudder as I noticed all this. I've never been on the "wrong" side of a lefty-green consensus before. This is how it looks. Except that it's not true, of course – this is not an issue that is divided by left or right, by green or anti-green, by climate change campaigners or deniers. The ideological reality is, like everything else I've touched along this road, far more subtle and complex than that.

I read through a list of questions and challenges on the ban from MEPs over the years – many are from Europe's right-wing parties, but not all. There have been queries and calls for clarity from all sides of the political spectrum. In Scotland, Labour MEP David Martin asked for evidence of the efficacy of the ban, while Conservative MEP Struan Stevenson highlighted health concerns.

In Westminster, Labour MP Sheila Gilmore brought the issue to an Adjournment Debate in 2012, raising the concerns of light-sensitive people and calling for "a dollop of common sense". It's noteworthy that she felt the need to begin by stating: "I do not seek to become part of a crusade by the *Daily Express* against the European Union." And she noted: "Some Honourable Members are raising their eyes towards the ceiling as I speak." But Conservative MP James Wharton congratulated her on securing this important debate: "There is also a whole spectrum of ways in which these bulbs can have a negative impact on the lives of our constituents."

In Switzerland in 2018, the Green President Regula Rytz teamed up with Yvette Estermann, the National Councillor for the conservative SVP, to contest the Single Lighting Regulation's proposed ban on halogen lighting: "What is the Federal Council doing to guarantee our citizens safe light for the school, for the office and for the home?" they asked.

As for the disparate, random bunch of us thrown together online to try to make sense of this story, we're from a wide mix of places and politics. I find it an interesting exercise in perspective shifting to walk a wide circle in my mind around the ban on incandescent lighting, viewing it through the different ideological lenses of people I've encountered. One was drawn to UKIP as it was the only party explicitly campaigning against the incandescent ban, and through raging frustration at the impossibility of reasoning with the EU. Another is a passionate left-winger, feeling that the fair treatment of minorities is a measure of any liberal democracy. Some are Green Party members and supporters agreeing that the weight of international law should be wielded to tackle climate change, the most vital issue of our time, but knowing that we've learnt a bit more about light in the last ten years. One is an anti-capitalist who sees the light bulb story as yet another facet of the corporate machine that is bulldozing every aspect of our society and sucking its lifeblood for profit.

We know it's irrelevant. We're all in this together. We might colour-code ideological perspectives, colour in blocks of our parliaments in red, blue, yellow or green. But light doesn't care. It does its stuff anyway: shining, scattering, refracting, zooming around the universe at 671,000,000 miles per hour. And ideology doesn't change who is more affected by light and who suffers from new forms of lighting. Some people simply have more sensitive skin, eyes, or nervous systems, or more finely tuned perceptions of flicker or frequency. But somehow ideology, language and symbolism have cast a veil over the debate, just as light pollution blurs the night sky. Somehow, light itself got lost along the way.

Once upon a time the incandescent light bulb was so vaunted that it became a symbol of a bright idea. The loss of this status was a challenge for illustrators and graphic designers, as a curly CFL just doesn't have the same illustrative impact for a "light-bulb moment". I've not seen anyone try an LED for this yet. For a while, the spiral twists of a CFL became a "green" symbol in cartoons and cloth carrier bags, but now CFLs are on their way out, leaving light bulb symbolism in a disjunctive limbo. Yet any discussion of light itself is inextricably embroiled with symbolism and metaphor. Light as good and dark as bad, light as modern and dark as old, light as safety and dark as danger. More light, brighter light, therefore, must be good, modern, safer . . . and more light . . . and yet more light.

When we think we're talking about light, especially light bulbs, we're really talking about ideology and symbolism and all the personal and cultural baggage that goes along with it. And while we're so caught up in an ideological tangle, light-sensitive people's lives are on hold. They will never get that time back again. It's time, now, to cut through the convoluted veil of politics, rhetoric and vested interests. How we got here is less relevant than where we go from here. It's time to appreciate and value darkness, and to truly talk about light.

9

Now What?

I sense two forces; both surging, growing, building momentum at astonishing speed and picking up power throughout the world.

One is a great wave, garish and roaring, thundering across the planet at a thrill-seeking pace. It's sparkling and dazzling, excited and chattering, calling in a shrill voice: "Brighter, newer, more, more, more." It grows and glows; you can watch its conquest from space – see the world fill up with smudges and stains and pinpricks of light as it spreads its shrieking brightness, vanquishing darkness all over our planet. It's driven by a voracious, insatiable monster called the economy, which harnesses technology, legislation and fervent belief as it pushes ever onwards, surging ahead with its delirium of newness, obliterating opposition with its sheer weight, pounding anything old, anything gone before.

This wave seems unstoppable. It's reaching tsunami proportions, surging and crashing and crushing, heedless to the devastation of all that lies in its path. And it would crush me, too, if it wasn't for another force that is also growing stronger, and is keeping me afloat, sustaining me.

Broader and quieter but no less powerful, a deep current is tugging just below the surface. It is swirling around the world, linking continents, carrying nutrients, surging ever onwards. It is stirring something ancient, reaching down to an ache from the dawn of time and carrying it forward. A longing for a purer, simpler way. It pulls and pulls, away from the shrill machinery of new, towards simplicity and authenticity. It speaks of things that are real, natural. Earth. Air. Soil. Fire. Sky. Things that have always been. Conversations by candlelight, digging a potato from the earth, the sensation of wind in hair. It craves real food, real light, real darkness. The movement of the moon and the enchantment of the stars. It seeks the night sky, rich and deep and mysterious, to be enclosed by it and opened by it to connection, with others, with the world and with something bigger, beyond us all.

Surfing the wave is a giddy acceleration of technological development that brings us new lighting in all shapes and sizes and colours and applications. It's created brighter towns and cities, dazzling headlights, light shining where it has only reflected before: on road signs, children's shoes, toys and gadgets. It illuminates through the day and the night. I've glimpsed things riding the wave that I scarcely know whether to believe . . . Plans for an artificial moon in the sky over a Chinese city? Satellite billboards selling advertising in space? Lasers in the headlights of supercars? It is hard to tell fact from science fiction anymore.

LED is so different from what has gone before that it is described as a revolution in lighting technology. It redefines the shape, form, content and very concept of artificial light. It has the potential to decouple "light" from "bulb". We can currently buy LED light for the home in a familiar, pear-shaped bulb to fit existing fittings, but this is a technology contorted to fit and mimic an incandescent legacy.

With LED, lighting technology becomes ever more intertwined with communications technology, moving from light for the purpose of illumination to a high-tech cog in the machinery of the Internet of Things. New LED street lights, for example, are being developed "smart city ready": with integrated antennae as part of their design. This enables them to host the roll-out of 5G – the fifth generation of mobile communication, which promises superfast connections within cities. 5G uses high-frequency wavelengths, which have a shorter range, and so it requires numerous antennae evenly spread over the target area. On top of concerns about light quality and the health effects of the electromagnetic frequencies involved, there are questions and protests about privacy infringe-ment and surveillance as street lights become continuous data collectors.

Not only is connectivity becoming embedded in light-ing technology, but the actual light itself is now being used as a medium to transmit data. The term "Li-Fi" was coined by Edinburgh University engineering professor Harald Haas in his 2011 TED talk, "Wireless data from every light bulb", in which he outlines a future symbiosis between lighting and communications technology whereby the presence of LED light enhances the potential for light data transmission.

Li-Fi, like Wi-Fi, uses electromagnetic waves to carry and communicate information, but whereas Wi-Fi uses radio frequency, Li-Fi uses the frequencies of visible light. Professor Haas explains that the radio waves used in Wi-Fi are limited in capacity and we're simply running out of spectrum to cope with the exponential growth in use. As this growth is only set to continue, we need to look else-where for capacity: to another part of the spectrum, such as light, which has 10,000 times more spectrum "space". Light, he says, is "inherently safe to use" because it's always been around us, indeed created us and created life. So wouldn't it be great to piggyback on the existing infrastructure of light, in the house and the street, and use that for wireless communications? There's one thing we need to do: "replace these inefficient incandescent light bulbs and fluorescent with the new technology of LED". This is because LED has the "cute property" that its intensity can be modulated at very high speeds, and Li-Fi works by encoding data in very subtle changes in brightness. Further developments, and a further TED talk in 2015, show how the solar cell can become a receiver for these high speed wireless signals encoded in light, offering the potential for the Internet of Things to develop without a corresponding surge in emis-sions. It seems that all that's needed, for all of this, is LEDs everywhere: in our homes, in our cars, in our streets and public buildings and transport systems. Everywhere.

Meanwhile, swelling in the deep current is a changing

attitude to, and renewed appreciation of, the night. And darkness itself – for so long shunned and feared and banished – is becoming welcomed and cherished, as more and more people in the over-lit Western world come to know the calm of it, feel the balm of it for an assaulted nervous system. And the night sky, taken for granted for too long, has become sought after and searched for – a focus of delight, discussion and attention, a tourist attraction in itself. "Astrotourism" is a growing trend, with travel tours geared towards stargazing and celestial activities such as the Northern Lights or a solar eclipse. There is also growing awareness in government – a new French law at the beginning of 2019 sets high standards for the protection of darkness. It is "designed to prevent, limit and reduce light pollution, including excessive disturbance to persons, fauna, flora or ecosystems, causing energy wastage or preventing observation of the night sky."

This movement is manifest in the work and growth of the International Dark-Sky Association (IDA). Based in Tucson, Arizona, the IDA was founded in 1988 by astronomers concerned about the loss of the night sky to light pollution. It has since grown into a global organisation with a holistic appreciation of the consequences of that loss, including the impact on the health of humans, wildlife and ecosystems. The IDA designates Dark Sky Places throughout the world, celebrating efforts to protect and promote the importance of dark skies. The designations include sanctuaries in the pristine skies of remote areas such as Stewart Island/Rakiura at the southern tip of New

Zealand, and Urban Night Sky Places, recognising planning and design that maximise the experience of night skies in the cities. Receiving these designations helps to strengthen protection by drawing attention to the night sky as a precious resource. It also means increased revenue from tourism, especially in the winter months. The Channel Island of Sark, without cars or street lights, was the first designation of a Dark Sky Island in 2011, and the Hebridean Island of Coll was the second in 2013. In 2018, Scotland celebrated its second Dark Sky Park, within the Cairngorms National Park, which followed the Galloway Forest Dark Sky Park in southwest Scotland.

The IDA recognises the importance of the night sky to wider human well-being, artistic inspiration and the sense of place in "our home galaxy" of the Milky Way. It awards recognition to individuals and "dark sky defenders" who have championed the night sky or campaigned against light pollution. These include artists and astronomers, scientists and students, filmmakers, lighting designers, and public organisers. I meet some of these defenders, online and in person – they are strong swimmers in the current. I reach out to them for strength and hope.

•

I'm fascinated by the relationship between individuals and organisations and the wider world. What interests me is how we are all caught up in vast tides of politics and power, whether we are aware of them or not. We're all swept along at random, sometimes carried on a warm current, sometimes churned and tossed in waves or snarled in passing debris. We're as powerless as microscopic plankton in the ocean. And yet, where do these great tides come from but the beliefs and imagination and determination of individuals? They grow and swell from individual minds to sweep across continents. We are astonishingly powerful and utterly vulnerable all at once.

For me personally, the story has now come not so much full circle as in a strange loop; there's a touch of irony that the future won't be fluorescent. It is now the turn of CFLs – the lights that brought me into this – to be phased out and banned by EU regulation. So recently vaunted as climate saviours, they will turn out to be just a blip in the history of artificial lighting.

"You can shut up now," say friends hopefully. "They'll soon be gone!"

I wish I could shut up – or celebrate. I would dance on their toxic grave but for the many reasons why I can't shut up, and why the CFL story still needs to be told within the ongoing and changing picture of artificial light.

For starters, CFLs are still with us and, even if only some of them last for their stated longevity, they will be for a long time – in homes and in hospitals, and still seemingly everywhere in toilets. There is still no effective labelling, for surely there has to be correlation between the language of a label and the scientific literacy of the intended recipient. How many people know that the little Hg sign in a corner of a light bulb box means mercury? (Because it used to be called by its Greek name, *hydrargyrum* – liquid silver). There is still little public guidance about keeping a safe distance from a lit bulb, or the importance of proper disposal for recycling spent or broken bulbs. CFLs are still being actively promoted throughout the world, regardless of whether there are any processes to cope with their disposal. The mercury they contain is still with us, and always will be, in some places accumulating in the water table and through the food chain.

Above all, the lessons from their story have not been learnt or even acknowledged, let alone applied to consequent legislation and technology. New products are still being rushed out with zeal, without sufficient testing or research, and health concerns are still being dismissed and ignored. And now the discussion about artificial light focuses exclusively on LEDs, CFLs as their immediate predecessors have become the point of comparison, and so LEDs are deemed "better" on almost every aspect.

"So are you okay with LEDs?" I'm often asked. *Yes and no*, I answer, or, *Sometimes, kind of, okay enough, in some ways*. Because in some ways, yes, the logistics of my own life are steadily getting easier as CFLs cede to LEDs in homes and shops, bus stations and concert halls. Some, especially when they're bright and blue, make me feel increasingly edgy and nauseous. I couldn't spend an evening under them but I can nip into a shop without lasting ill effect. Some, softer warmer ones, I find quite pleasant. For me personally, physiologically, LEDs are indeed better than CFLs, if that's all we have to compare them with.

So am I okay with LEDs? Intellectually, no, because

the questions of the impact on many aspects of human health have not yet been answered. Morally, absolutely no, because the suffering of those who cannot tolerate LED light is simply being ignored. No, I'm not okay with the mandatory use of a new technology that excludes some citizens from participating in civic life. The wave thunders on, and I can hear the cries of those in its path, whose life chances are being shattered. I hear it every day on social media: the fear, the desperation and excruciating pain. It's getting louder.

Oblivious, there's a triumphant tone to lighting discourse: a widespread enthusiasm from some lighting designers, museums and galleries, and street light engineers, that we've found what we were looking for in LEDs. The message from all sides seems to be: "It's okay, we've got this – it's all good now. Yup, CFLs were rubbish really. Yep, early LEDs were harsh, too bright and too blue, but we're sorted now. Colour, softness, astonishing versatility – we can make them do whatever you want them to, and they're still getting better and better." It's a "win-win" – just as many proclaimed in the early days of CFLs. Even a "win-win-win-win" for maintenance, cost, environment, flexibility.

Who'd want to know about the losers?

•

As the new generation of LED lighting settles in and develops further, a curious thing has happened to the dialogue of artificial light. Once upon a time, health and human well-being were largely absent from the discussion when advertising and marketing lighting (aside from the marketing of "SAD lamps" – a name which always gives me a wry smile – for the bright lights marketed for seasonal affective disorder). Then all of a sudden it became a hot topic: "human centric lighting", "circadian lighting", "biological lighting", "light for health" – whatever you want to call it, it seems to be all the rage.

Human centric lighting is defined by the industry body LightingEurope as lighting that "supports the health, well-being and performance of humans by combining visual, biological and emotional benefits of light". It may be "the biggest jump since the electric light bulb", says the Human Centric Lighting Society, amid palpable excitement that LEDs – or "solid-state lighting" – offer the solution, not just to issues of energy efficiency but to health and well-being. It's already being tried and applied in offices, hospitals and schools.

The idea is that this new concept and technology will revolutionise our relationship with artificial light – by acknowledging the power of light on physiology and psychology and enabling us to harness that power and manipulate it for the desired effect. With a remote control and smart light source, the colour and brightness of artificial light can be adjusted whenever needed to mimic daylight and its corresponding response in a human. Simply up the brightness and blue to tell your body and mind that it's morning and so to feel more alert and active; programme it in advance to wake you with the appropriate colour temperature whatever the time. Soften and warm the light to wind down and get that melatonin flowing prior to sleep.

Except that it isn't so simple. Light is far, far more than colour temperature and brightness, and the cues that our bodies and minds take from it are far more subtle and complex. Are we still messing with something that we just don't really understand? Some lighting designers and medics fear that we are, and point to insufficient research in this field to substantiate the many excited claims. Advances in our understanding of light and life have been comparatively recent, especially in ecology and neurology. Take the discovery of the role of the third photoreceptors in the eye – those intrinsically photosensitive retinal ganglion cells (ipRGCs) – which has changed the understanding of the relationship between light and the brain (see Chapter Four). What else have we yet to discover?

"We can't develop human centric lighting until we know what impact light has upon human biology across the day and night cycle," states neuroscientist Professor Russell Foster from Oxford University, who led the team that discovered ipRGCs.

"Are we as lighting designers practising medicine without a licence?" asks architect and designer Karolina Zielinska-Dabkowska. She warns of widespread misunderstanding of the very nature of LED lighting, explaining that its unique characteristics directly conflict with the aims of human centric lighting. The stated *raison d'être* of the LED lighting revolution is efficiency – saving energy to mitigate global warming – yet the most energy-efficient produces blue-rich white light, which has serious effects on the environment and human health. Whatever is then done to that light to render it warmer, and more akin to incandescent lighting, increases the cost and reduces the efficiency.

Much of the discussion and debate focuses on colour temperature (see Chapter Three). Karolina feels this is a limited and insufficient way to describe light. With LEDs this number only indicates an approximation of the equivalent colour of "real light" that it is effectively mimicking. So a "daylight" LED may look like the natural bright light of midday but the spectrum of that light is very different to sunlight. LED light lacks infrared, and the smooth, steady curve of natural light. It is the actual frequencies of light that have the most profound biological effects, and so, Karolina argues, we need to focus on the spectral power distribution – which wavelengths of light are present in what intensity – more than apparent colour.

As I browse the claims – "enhance mood", "stay awake and alert", "relax" – I can't help thinking that we did all that already, with a piece of technology called a dimmer switch. In our home nearly all our light fittings have dimmer switches and I use them to alter the mood according to activity. Bright and upbeat around the table for family activities in the evenings or intricate Lego creations in the children's bedroom, then softened a bit for dinner or storytime. Just a tiny twist for a hint of light for sleepy eyes waking up in the night, or coming round on a winter's morning.

Now What?

Concurrent to so-called human centric lighting, there has been a growing surge of "eco" lighting, designed and adapted to minimise impact on wildlife. In 2018, a new development of housing beside a nature reserve in the Netherlands installed "bat-friendly lighting" in what was celebrated as a "world first", using a system of red LED lights that "is perceived by bats as darkness" and so doesn't interfere with their internal compass.

·

Light pollution, human centric lighting, circadian rhythms, Li-Fi and lasers, ecologically sensitive lighting, bat-friendly lighting . . . Suddenly, ironically, there seems to be an awful lot of talk about light. But it hasn't fully filtered through to the general public yet, especially in the discourse about health and well-being. I flick through magazines in a halogen-lit dentist's waiting room. There are endless articles about diet, exercise, stress, lifestyle, mental health, posture . . . but nothing about light.

My daughter at primary school learns about Basic Human Needs: "People need food, shelter and water," she tells me solemnly. There is some discussion about whether access to the internet now constitutes a basic human need. There is no mention of access to healthy light.

It will break through soon. Within specialist fields such as lighting design, architecture, environmental policy, it is growing from a mutter to a babble and a shout. And yet this is happening with a restricted vocabulary. Something is missing, conspicuous in its absence from all these concepts and conversations, conferences and webinars: incandescent.

As incandescent bulbs disappear from shops and sitting rooms, the tools to manufacture them dismantled and gone from the factories, the very word "incandescent" is becoming lost from the dialogue. Is it a taboo subject, or has it been forgotten so soon? As if "progress" is a juggernaut travelling in just one direction and has already left incandescent lighting far behind. The focus of future discussion is LEDs and their related technology. Law, technology and language have conspired to decide what is the future and it then becomes an accepted fact – consensus congeals around it.

It's as if only LED exists now as an option. Take street lighting, for example. The possibility of continued use of sodium lights doesn't enter the debate. Even discussion about astronomy and ecology is phrased in terms of recommendations for LEDs: they are to be shielded, directed down, to have a low colour temperature, soft and red-rich. The questions are phrased as: "What can we do to make LED light as good as possible for astronomy, for fast-flying bats?" rather than: "Is this the best technology for lighting our streets?"

Has anyone else noticed that this seemingly inevitable future direction is a choice that has been made? It was a political decision – actually, a series of political decisions by those tasked with making decisions on our behalf – not some immutable law of physics or natural progress of evolution. At the time of these decisions, the European

Now What?

Energy Commissioner, Günther Oettinger, acknowledged the potential problems for light-sensitive people, but said it was "unrealistic" to go back to incandescents. Instead, he assured campaigners, resources and funds would be channelled into research and further technological development in a quest to find healthier light.

Now, more than ten years on, amidst growing disquiet about the negative effects of LEDs on human and environmental health, that quest continues. There's a sense that the "right light" is still out there in the future and we have to fund and facilitate a path towards it. A 2018 lighting magazine suggested a Nobel Prize should be awarded "to those who discover the key to healthy artificial lighting". Perhaps the key is not in the future, but mistakenly discarded into the past.

As Jane Brox, author of *Brilliant: The Evolution of Artificial Light*, says: "Such stubborn fondness for the age of incandescence is more than simply nostalgia. It's testimony to how much incandescent light has meant, and how perfectly suited it still seems to be, to modern life: the steady, brilliant light of a speeding century: light born of invention but also warm . . . 'old-fashioned' bulbs still shed a more satisfactory light than anything yet developed to replace them. And perhaps they always will."

10

Reflections and Refractions

You hear of light dancing on the water's surface, but this is bursting out over the banks and having a veritable boogie with the surrounding willows: shimmying up their trunks in flickering hula hoops, disco shimmering all over the silver-green leaves. I'm sitting on the bank of the River Cam, watching the constant stream of boats, bikes, birds, dog-walkers and joggers all passing through this riverside light show. Mute swans glide by, leaving streaks of golden wake through the splintered patterns of willows and sunshine.

Since I started writing this book, one of life's freak waves snatched me away from Scotland and landed me in Cambridge, right beside the river. I'm still reeling, adjusting, absorbing the newness. I wasn't sure what I'd make of such flat land, especially with a Dutch husband who had settled in Scotland because he loved mountains and the wild. Yet wildness ripples through this small city, in owl calls at night and glimpses of water voles at the river's edge, in peregrine falcons nesting among the city spires. We watched a peregrine hover and stoop, dropping vertically through the sky as if plunging down an invisible lift shaft. Out of town the light is vast. I realised – driving towards Norfolk for a weekend – that one of the things I've always loved about being at sea is the constant expanse of uninterrupted sky.

On the first week here I was lying in the bath when the frosted glass window of the bathroom lit up with sudden, juddering light. Then again, then again. I grabbed a towel and peeked out, wondering if it was the emergency services and why there were no sirens. Bike lights! The cycleway runs beside our home and many of the bikes have astonishingly bright front lights that flicker and shudder and dazzle. I'd thought cycling along the river would be a pleasant way into town after dark, avoiding the bleak fluorescent street lights on the minor roads. But no, the river is busy with bikes at all hours, their front lights louder than car headlights, often flashing with a violent, throbbing pulse. I want to twist away and face only the darkness to my left, to shield or shut my eyes, but none of this is conducive to cycling forwards. I've learnt to pick my route along the main roads now as they're lined with gentle, amber street lights.

Otherwise the night sky is darker than I'm accustomed to, having lived for a decade near the bright sky glow from a huge oil refinery. Walk the common land beside the Cam at night and you can see the stars, pick out the main constellations. On Wednesdays, Cambridge University Observatory is open so the public can access the powerful telescopes and the expertise and enthusiasm of students and staff. My daughter's eyes and mouth are wide circles, her whole face an "O" of wonder to see the rings of Saturn. My son is on first-name terms with distant nebulae. We learn that when we look up we are actually looking back in time and see bits of the universe's history. Because of the time it takes for light to travel, what we are seeing is light that has already happened.

What we see with our own eyes is our "neighbourhood" in astronomical terms, maybe tens of thousands of years old, which is not old in terms of star-life. I like the idea of having a solar neighbourhood, within ten light years away, some stars "just" three to four light years away. Some of the brightest stars are a few hundred light years away. What we see of the Milky Way is, on average, a few thousand light years away. With a sensitive telescope you can look back billions of years in time almost to the beginning of the universe.

We learn that if you shine a torch away from you in space you can't see the beam, because we don't actually see light unless it's directed at us, just what it encounters and reveals to us. So what we think of as a torch beam is actually light hitting bits of dust and stuff in the air. That's why space looks so dark. There are very few molecules floating out there. The sun emits a huge amount of light but we see the sky as blue because a tiny bit of light occasionally hits these molecules and scatters, which appears to us as blue. We stare up into the sky. It's smudged soft orange from the nearby city lights, but we can see the main constellations clearly with our eyes, and the details of the moon's surface through the telescopes, all shadows and ragged-edged craters.

Deeper darkness is in easy reach beyond the city: Norfolk boasts a number of Dark Sky Discovery sites and draws stargazers to its clear night skies. We head out for weekends when we can. And each time brings the reminder that real darkness is surprisingly un-dark – the sky out here feels alive with bright clusters and shapes and patterns of starlight, shot with sudden shooting stars or washed with silvery moonlight. I sleep with the campervan window open beside me, delighted if I wake in the night and can feast on the sky.

Once upon a time I was scared of the dark. As a child I'd walk my dog along the road to where the village street lights stopped. Beyond there was darkness, enticing, mysterious and terrifying; I'd peer into it, but didn't have the courage to step further than the last pool of orange light. All my adult life I've slept with a light on if I'm alone. Whatever was I scared of? I wonder now. Whatever my imagination spooked me with, in all its fears and worries and nightmares, I never could have foreseen that the thing I'd actually have to worry about would be light itself.

Then along came this strange story about light, which changed my life, and changed me as it moved steadily outwards in concentric circles. First there was me, in my own little world, suddenly feeling *bleurgh* when I encountered curly light bulbs. Then I found it was bigger than me: that people all over the world were getting ill from new forms of light. And it was bigger than CFLs: other new lights such as LEDs were causing worse problems for more people. Then unanswered questions grew and bulged, asking how these lights were affecting everyone's eyes and skin and brains. And, way beyond bodies, light expanded into a global swirl of ideas, symbolism, ideology, profit and power. Way beyond people, artificial light was altering the clock that beats throughout the natural world. And everything's relationship with all that's beyond our world – the night sky, the stars, the Milky Way, the universe. Light *was* insubstantial, irrelevant, nothing really, just a shield from the dark; now, light is physical, tangible, acutely relevant. Everything.

I love the dark nowadays. I feel a physical longing for its depths and texture. I seek it out, plan my time around it, organise my trips to it like a pilgrimage. Though I confess I still prefer to encounter it with at least a dog for company. I still freak sometimes in a blacked-out room; I wake disorientated and panicky if there are shutters or thick curtains and blinds. I seem to need constant sky, night and day, a glimpse of it at least. But I treasure real dark, especially when it's sprinkled with faraway light from thousands of light years away.

•

I still pinch myself sometimes: to think that this really happened . . . is happening. Incandescent light is being purged from pretty much the whole wide world. A simple, harmless, beautiful household product was demonised and obsoleted, then deleted, deemed illegal because . . . because it uses electricity? And so the world thunders on without it, ever brighter, lighter, techier, wired up and wireless and always connected and using ever more electricity each and every moment. And the loss of incandescent beauty is glossed over, already forgotten and replaced with apparent equivalents. Except that there is no equivalent. I believe all of our built environment is poorer without it, and some people will be unable to live a normal life without it. I guard my remaining stock carefully, flitting around the house switching lights off to preserve the last treasures.

My incredulity cedes to a cold wave of despair and then, again, anger begins to burn. Sometimes a helpless, ragged rage; sometimes I want to blame something. Who or what is to blame for this colossal, gaping, global loss of something essential and irreplaceable?

Is it the lighting manufacturers, for encouraging the replacement of a patent-expired technology with a more profitable alternative?

Or is it the governments of the world, for rushing to legislate in a token gesture, without proper scrutiny or democratic process or debate, in a bid to be seen to be green whilst bowing to corporate interests? Or was this a genuine attempt to use the power of legislation to seize one small opportunity to help mitigate the urgent issue of climate change?

Or is it the environmentalists, including individual politicians, parties and organisations, for thrusting the changes forward before the technology was ready or tested, grasping at figures which suited their goals without querying their provenance or consequences or even truth? Or was this a genuine belief in a chance to do something real –

that one small change extrapolated over the world would have a significant effect? Was it a sense of a lesser evil – the way tough decisions have to be made in the urgency of environmental demands today, like backing even the environmental monstrosity of nuclear power in the face of the catastrophic prospects of climate change?

Is it the media, for failing in its fundamental duty to scrutinise policy and the motives of those in power, for the grave, global deficiencies in scepticism, and for wielding so glibly the powerful tool of language? Or did they simply not know what was really going on? Because we don't, as a society in general, talk about or know about light.

Is it the public, for barely noticing, for believing what they were told and scarcely questioning, for so quickly accepting the consensus? People like me, I guess, trying so hard to be Good. And busy – going to work, getting the dinner on, watching gannets. How many of us would plough through the appendices of weighty tomes of EU regulation to question the finer detail? Who would query, dissect or check figures from passionate environmentalists who dedicate their working lives to the care of our planet? I mean, why would you?

Ultimately, I blame ignorance: a gross underestimation, from all the above, as to the power and importance of light and the complexity of its interactions with life on Earth.

Living this story has changed not just my perception of light and dark, but of politics and power and science. I've become sceptical of all "eco" claims, even the very word, and I'm constantly alert to anything that might be greenwash. To some extent you have to trust – you can't question everything – but who do you trust?

When we were re-doing our bathroom we wandered around B&Q, looking at toilets. We chose the one that said "eco" – I couldn't tell you what made it so . . . less water, I guess? I didn't have the time or energy to scrutinise the claim, and just had to trust that someone somewhere is checking this labelling. I felt a little bit better when I flushed the toilet that something relatively eco was going on. Is that what the legislators did, dashing from meeting to meeting for their various committees, maybe dealing with lights in the morning and washing machines in the afternoon? "Ah, this one says 'eco', so let's go with that."

When my light troubles began, I wrote to a friend who is a prominent environmental activist. He said he didn't know whether the green claims about CFLs were true, but the consensus was that they are. I felt dismissed, my question crushed by this heavy, solid thing called consensus. I didn't realise until years later that he'd hit the nail on the head. The consensus: if I've learnt anything from my part in this strange story, it's about the nature of, and danger of, consensus. How fast we accept notions of good and bad when those around us do. How easy it is to accept what we're told uncritically, without questioning. How consensus can solidify into something implacable, how uncomfortable it is for that to be challenged, and how threatening are those who challenge it.

The consensus is entwined with our attitude to science. "Scientists say" can be enough of an explanation in itself to designate something as accepted fact. Science is a process of discovery, constant and ongoing, of revisions and rejections of what was previously accepted as fact, of a perpetual questioning. "The science" is simply where we've got to along that journey in any given moment. "There is no evidence" can mean no research has been done on a certain subject. And when research has been carried out, we need to ask how and by what methodology and who funded it.

I'm pondering all this while clearing breakfast dishes one bright morning in our Cambridge kitchen when BBC Radio 4's *National Health Stories* begins to tell of an epidemic of lung cancer in the early days of the NHS. The resonance makes me put down my dishcloth and sit down to listen. Was chimney smoke causing such widespread lung cancer, people wondered, or was it motor cars? Or carcinogenic substances in tar? Smoking was not a prime suspect, although eighty per cent of men smoked: "It was difficult to believe that this apparently innocuous habit . . . could be harmful." Most doctors smoked, and sometimes even prescribed cigarettes to their patients to clear their lungs. Then, in 1954, a study of 40,000 doctors confirmed the link between smoking and lung cancer. It was not news that the public or the government wanted to hear, and the health minister chain-smoked throughout the press conference to announce the results. Part of the disbelief was linked to the new concept that causality operated over decades: that you could start smoking in early adult life and then get lung cancer in middle or old age. The public were hard to persuade, as was the media, and indeed the government "who were more convinced by the tobacco industry than by their own top doctor". The Permanent Secretary was well connected to tobacco companies, and the tobacco industry dismissed a damning report on smoking by the Royal College of Physicians. It took more than thirty years for the British public to take it seriously, even though we had the worst death rate from lung cancer in the world.

It's impossible to hear this story without drawing parallels to light. Can we begin again, and this time talk about *actual* light? Can we strip away the tangled mesh of language, symbolism and ideology and clearly see the science beneath? What do we know about light and life and human health and the impact on the environment? Can we bring that all to the table and look anew at light, and energy use and the severity of climate change. How much artificial light do we really need? What's the best way to meet that need? What is the best technology to provide the actual light that is required?

•

I stroll around the buildings of the Cavendish Laboratory of Cambridge University. There are a lot of people here who know an awful lot about light, and mind-blowing research is taking place in the various departments – NanoPhotonics, the study and harnessing of light on the scale of a billionth of a metre. There's Optoelectronics, and across the road is Astrophysics, and Quantum Matter. Here, you can focus, bend and scrutinise light, study it across great galaxies or in quantum detail. But you won't learn here about light's interaction with the body and mind, let alone with the soul.

Over on the other side of the city, there's the huge Addenbrooke's teaching hospital, which boasts world-class research. It has departments of Neurosciences, Ophthalmology, Dermatology, Psychiatry . . . but I wonder what you could learn there about the interaction of light with all this, how new forms of artificial light differ in their dynamics with photons and tissue, neurotransmitters and hormones?

I pass the rainbow-coloured frontage of the Maxwell Centre and I wonder if we need a new Maxwell. What struck me about hearing James Clerk Maxwell's story is how he had observed and questioned and learnt so much about electricity, but then teamed up with experimental physicist Michael Faraday, who had come to understand so much about magnetism. Pooling their resources, their combined knowledge of electricity and of magnetism, led to the great leap forward in discovering electromagnetic radiation. Maybe it's time for another partnership and another breakthrough, this time between physics and physiology. We need a gigantic leap forward in understanding how that whole spectrum of electromagnetic radiation – including the small sliver of it that we call visible light – interacts with life. And not just with human life, but the trees and bats and birds and insects too, with all of life on Earth. And still people will scoff, as they did in Maxwell's day that something that cannot be seen is having a very real effect on the building blocks of matter that constitute life.

And one day we will come to accept, respect and understand that this stuff is real, so very real, whether visible or not.

And, meantime, while we wait for this big breakthrough in scientific understanding, I've come to realise that we already have something that understands this relationship precisely. We have something with the capacity to detect changes in light on a cellular level, and it doesn't need any specialised equipment in a laboratory. It's been developed in nature's own Research and Development department over millions of years: it's our own bodies and brains, our cells, our selves.

It lets you know. So listen. Listen carefully and feel it. It may not be an extreme reaction of blinding pain or debilitating symptoms. Listen to the subtle stuff: a scratchy irritation; a gut feeling of "I don't want to be here" or "Ooooh, I like this place"; a cringe, a wince, or a boost in your energy levels. It's all your body, detecting light with exquisite precision, and letting you know if it's right for you.

I believe there is a Goldilocks light for each and every one of us at any moment in time: not too bright, not too dim, not too red and not too blue. Just right. You can feel it, maybe not articulate it or know intellectually what it would be, but your body and mind recognise it instantly, and respond with a softening, a little surge of joy and relief when you encounter light that is just right.

For me, that is daylight wherever possible, and I'd still choose the light of a Scottish coastline if I could. I'd choose firelight for deeper conversations; candles for intimacy. As for artificial light, when it's necessary, my Goldilocks light is a 60 watt incandescent bulb on a dimmer switch: soft and gentle, then bright, warm and uplifting.

How about you?

Glossary

age-related macular degeneration (AMD) – an eye condition that affects the central part of the retina (the macula) and damages central vision.

blue light – the high energy range of visible light on the *electromagnetic spectrum.*

circadian rhythm – physiological processes in an organism that follow a daily cycle. Also known as "body clock".

compact fluorescent light (CFL) – a *fluorescent lamp* that is a tube curled or folded into roughly the same size and shape of a conventional light bulb.

correlated colour temperature (CCT) – the colour appearance of light emitted by a lamp, based on the colour of light from a reference source when heated to a particular temperature, measured in degrees *Kelvin (K)*. The CCT rating for a lamp is a general "warmth" or "coolness" measure of its appearance, with a higher CCT appearing "cooler".

Ecodesign Directive – an EU framework to lower the energy consumption of electronic products by setting minimum mandatory requirements for energy efficiency.

electromagnetic field (EMF) – a combination of electric and magnetic fields of force, produced by the generation and use of electricity.

electromagnetic radiation – the waves (or their quanta/*photons*) of the *electromagnetic field* propagating through space and carrying electromagnetic radiant energy.

Glossary

electromagnetic spectrum – the entire range of wavelengths of all known *electromagnetic radiation*, including radio waves, microwaves, infrared, visible light, *ultraviolet*, X-rays and gamma rays. See also *visible spectrum*.

flicker – rapid, repeated changes in the brightness of light.

fluorescent lamp – an electric light that is normally a tube coated on the inside with fluorescent material, in which *mercury* vapour produces *ultraviolet* light, causing the coating to emit visible light.

halogen lamp – an *incandescent light* source that is filled with a mixture of an inert gas and a small amount of a halogen (a group of chemically related elements, including chlorine and iodine) that recycles the *tungsten* back to the filament.

human centric lighting (HCL) – lighting claimed to support the health, well-being and performance of humans.

incandescent light – an electric lamp that works by incandescence: heating a wire filament to a high temperature so that it emits light. See also *tungsten*.

International Dark-Sky Association (IDA) – a US-based organisation originally founded by astronomers to advocate for the protection of the night sky. It now spearheads a global movement against light pollution.

intrinsically photosensitive retinal ganglion cells (ipRGCs) – specialised cells in the retina of the eye of mammals that respond to light separately from the rods and cones. They play a major role in synchronising the *circadian rhythm* to the 24-hour light/dark cycle.

Kelvin (K) – a measure of temperature in degrees (like Fahrenheit or Celsius), which uses "absolute zero" as its lowest point. Kelvin is used as a scale of *correlated colour temperature (CCT)*.

Li-Fi – wireless technology that uses visible light to carry and communicate information data.

light-emitting diode (LED) – a semiconductor light source that emits light when an electric current flows through it. Also known as "solid state lighting".

light sensitivity – an intolerance of light, whether light in general or certain aspects such as brightness, intensity, colour temperature.

lumen (lm) – a measurement in units of the amount of visible light emitted from a source.

lux – the unit of illuminance, equal to one *lumen* per square metre, measuring the amount of visible light on a surface area.

melanopsin – a type of photopigment in specialist eye cells that is involved in regulating *melatonin.*

melatonin – a hormone that regulates sleep–wake cycles and influences the *circadian rhythm* of an animal (including humans).

mercury – a chemical element (Hg), also known as quicksilver, used in a *fluorescent lamp.* Mercury and its compounds are highly toxic to humans and the environment.

metal halide lamp – an electric light that uses *mercury* vapour combined with metal halides (compounds between metals and halogens) to create a powerful light. It is a High Intensity Discharge (HID) light, meaning that it creates light from the electric arc within a small discharge tube.

nanometre – a very small metric unit of length (one millionth of a millimetre), used to express the wavelengths of light.

photon – the smallest specific amount (or quantum) of energy in *electromagnetic radiation.*

Scientific Committee on Emerging and Newly Identified Health Risks (SCENHIR) – an EU advisory committee active between 2004 and 2016 that was tasked with identifying and assessing potential risks to human health and the environment.

Scientific Committee on Health, Environmental and Emerging Risks (SCHEER) – the EU Committee that succeeded *SCENHIR* in 2016 and covers risks to consumer safety and public health.

seasonal affective disorder (SAD) – a form of depression thought to be linked to lack of sunlight and thus follows a seasonal pattern. Sometimes described as "winter depression".

Single Lighting Regulation (SLR) – an EU ruling intended to streamline the different regulations for lights that resulted from the *Ecodesign Directive.* It aims to update terminology and further increase efficiency.

sodium lighting – a form of electric light that works by creating an electric arc through vaporised sodium metal. Traditionally used for street lighting in two forms: low pressure (LPS) and high pressure (HPS).

tungsten – a rare metal and chemical element that can be heated to a very high temperature. It is used for the filament in *incandescent light* bulbs, also known as tungsten lamps.

ultraviolet radiation (UV or UVR) – a higher energy band of the *electromagnetic spectrum* that is invisible to humans (though some birds and insects can see it). UV radiation is divided into three wavelength ranges, with increasing energy (and so potential harm): UVA, UVB and UVC.

visible spectrum – the range of wavelengths of *electromagnetic radiation* that human eyes perceive as visible light (from red to violet, as seen in a rainbow).

watt (W) – the unit of power that specifies the rate of energy transfer, such as the consumption of electricity by a product.

World Health Organisation (WHO) – an agency of the United Nations concerned with promoting and protecting public health.

Notes and References

Chapter 1

page
12 Canadian blog about the effect of CFLs: https://cfls.wordpress.com/

13 "Getting rid of incandescents is a no-brainer": "WWF urges EU to ban all energy wasting light bulbs", WWF European Policy Office, 28 August 2009

13 WWF called for the shops to be cleared of incandescents: "Light bulb ban begins today", WWF press release, 1 September 2009,

13 WWF called for EU anti-dumping duties to be relaxed: "EU keeps unfair market barriers on energy-saving lamps", WWF European Policy Office, 29 August 2007

13 Greenpeace campaign to promote CFLs in India: "Greenpeace calls India bulb firms 'climate villains'", Reuters Environment, 30 April 2007

13 Greenpeace crushed 10,000 bulbs: "10,000 energy wasting lightbulbs crushed", Greenpeace East Asia website, 21 April 2007

15 Each CFL bulb contains mercury: https://www.epa.gov/mercury/health-effects-exposures-mercury

16 The average CFL bulb in India contains 21mg of mercury: *Toxics In That Glow – Mercury in Compact Fluorescent Lamps (CFLs) in India*, Toxics Link, 2011

16 NHS guidance in event of a CFL bulb breaking: https://www.nhs.uk/common-health-questions/accidents-first-aid-and-treatments/

Notes and References

can-a-broken-thermometer-or-light-bulb-cause-mercury-
poisoning/#light-bulbs

16 US EPA guidance in event of a CFL bulb breaking: https://www.
epa.gov/cfl/cleaning-broken-cfl

16 Most CFL bulbs are not disposed of properly, Toxics Link: *Toxics
In That Glow – Mercury in Compact Fluorescent Lamps (CFLs) in
India*, Toxics Link, 2011

21 Government guidelines for using CFLs: "Emissions from compact
fluorescent lights", Health Protection Agency, 9 October 2008

28–29 Winter survival in Finland: Erkkilä, Jaana Helvi Maria,
"About Time and Light" in *Surprised by Words – An Anthology*,
Grosvenor House Publishing Ltd, 2012

Chapter 2

page

47–48 Employment tribunal ruling report: *Private Eye*, 26 October
2008, p.31

Chapter 4

page

83 Professor John Hawk at a meeting about CFLs in Brussels:
extract from correspondence with the Spectrum Alliance

84–85 Cracks in the phosphor coatings of CFL bulbs: Mironava,
Tatsiana, "The Effects of UV Emission from Compact Fluorescent
Light Exposure on Human Dermal Fibroblasts and Keratinocytes
In Vitro", *Photochemistry and Photobiology*, 2012

85 British Association of Dermatologists' statement on CFLs:
"Ultraviolet (UV) emissions and Compact Fluorescent Lights",
British Association of Dermatologists Group Position Statements,
January 2013

85–86 Dundee University Hospital study into the effects of
CFLs: Fenton *et al*, "Energy-saving lamps and their impact
on photosensitive and normal individuals", *British Journal of
Dermatology*, 2013

86 Official UK government advice for using CFLs: "Emissions from

compact fluorescent lights", Health Protection Agency, 9 October 2008

86 "Some very sensitive people cannot tolerate" LEDs, Lupus UK: https://www.lupusuk.org.uk/eclipse/

89 Description of life with seborrhoeic eczema: Lyndsey, Anna, *Girl in the Dark*, Bloomsbury, 2015

90–91 Borek Puza on the dramatic increase in light intensity: https://lightmareaustralia.weebly.com

91–92 French government report on health and LED lights: "Opinion of the French Agency for Food, Environmental and Occupational Health & Safety in response to the internally-solicited request entitled 'Health effects of lighting systems using light-emitting diodes (LEDs)'", ANSES, 19 October 2010

92 Warning about prolonged exposure to LED lighting: Chamorro, E., *et al*, "Effects of light-emitting diode radiations on human retinal pigment epithelial cells in vitro." *Photochemistry and Photobiology*, 2013

92–93 "Retinal damage and cell death in rats": Lougheed, Tim, "Hidden Blue Hazard? LED lighting and Retinal Damage in Rats", *Environmental Health Perspectives*, Volume 122, March 2014

93 "Neuronal cells are incapable of repairing themselves", Professor Chang-Ho-Yang: *ibid*

93–94 RAC survey: "New car headlights present 'unwanted safety risk' to drivers", *RAC News*, 25 March 2018

94 *Goodwood* columnist on driving cars with strong LED lights: Simister, John, "Why LED headlights are dangerous", *Goodwood Road & Racing*, 16 April 2018

94 Professor John Marshall stockpiling bulbs: Naish, John, "The medical experts who refuse to use low-energy lightbulbs in the home", *Daily Mail*, 13 May 2014

94 Professor John Marshall concerned about "blue and UV biologically unfriendly sources": Brogan, Rory, "Professor John Marshall warns of low-energy bulb risks", *Optician*, 16 May 2014

94–96 Professor John Marshall's series of opinion pieces: Marshall, John, "Understanding the risks of phototoxicity on the eye",

International Review of Ophthalmic Optics, October 2014; and "The blue light paradox: problem or panacea", *International Review of Ophthalmic Optics*, February 2017

96–97 Light sensitivity is common among migraineurs: "Migraine and Light Sensitivity", Migraine Action, 2010–2011

97 Recent research shows a complex relationship between migraine and light: Noseda, R., *et al*, "Neural mechanism for hypothalamic-mediated autonomic responses to light during migraine", *Proceedings of the National Academy of Sciences of the United States of America*, July 2017

100–101 Canadian investigation into emissions from CFLs: "Shedding some light on compact fluorescent bulbs", *16x9 – The Bigger Picture*, Global News, Canada, 24 August 2012

106 Harvard study that compared the effect of exposure to green and blue light: "Blue light has a dark side", *Harvard Health Letter*, May 2012 (updated 13 August 2018)

107 Shift work is "probably carcinogenic", International Agency for Research on Cancer; and, Recent studies have shown a significant increase in the risk of breast and prostate cancer: Garcia-Saenz, A., *et al*, "Evaluating the Association between Artificial Light-at-Night Exposure and Breast and Prostate Cancer Risk in Spain", *Environmental Health Perspectives*, April 2018

107 Certain drugs used to treat breast cancer are rendered less effective: Dauchy, R.T., *et al*, "Circadian and Melatonin Disruption by Exposure to Light at Night Drives Intrinsic Resistance to Tamoxifen Therapy in Breast Cancer", *Cancer Research*, August 2014

107–108 Study that analysed data from over 100,000 UK women: Johns, L.E., *et al*, "Domestic light at night and breast cancer risk: a prospective analysis of 105,000 UK women in the Generations Study", *British Journal of Cancer*, 118, February 2018

108 *Which?* report on SAD lamps: https://www.which.co.uk/reviews/sad-lamps/article/sad-lamps-and-seasonal-affective-disorder/choosing-a-sad-lamp

108 NHS guidelines on light therapy: https://www.nhs.uk/conditions/seasonal-affective-disorder-sad/treatment/

Notes and References

109   Study into quality of light and influence on emotions:
       Vandewalle, G., *et al*, "Spectral quality of light modulates
       emotional brain responses in humans", *Proceedings of the
       National Academy of Sciences of the United States of America*,
       November 2010

109–110 The direct role of light on mood and cognitive functions:
       Hattar, Samer, "How light affects our minds, our moods and our
       sleep", *Braintalk* podcast, Johns Hopkins Brain Science Institute,
       4 April 2013

110   Red and infrared light can have important protective effects on
       tissues and organs: Sivapathasuntharam, Chrishne, *et al*, "Aging
       retinal function is improved by near infrared light (670nm) that
       is associated with corrected mitochondrial decline", *Neurobiology
       of Aging*, April 2017

111   Fluorescent light could be a risk factor for people with certain
       skin conditions: report on light sensitivity by the Scientific
       Committee on Emerging and Newly Identified Health Risks
       (SCENHIR) of the European Commission, 23 September 2008

111–112 "Health Effects of Artificial Light": report by the Scientific
       Committee on Emerging and Newly Identified Health Risks
       (SCENHIR) of the European Commission, 19 March 2012

112–113 Potential health effects of LED lighting: report by the
       Scientific Committee on Health, Environmental and Emerging
       Risks (SCHEER) of the European Commission, 6 June 2018

Chapter 5

page
122   Scientists measured the brightness of artificial light at night:
       Falchi, F., *et al*, "The new world atlas of artificial night sky
       brightness", *Science Advances*, June 2016

124   NASA data of the Earth at night: https://earthobservatory.nasa.
       gov/features/Malvinas

125–126 "Language of light" in the ocean, Edith Widder: https://www.
       ted.com/talks/edith_widder_the_weird_and_wonderful_world_
       of_bioluminescence

126 Birdlife International recognises light pollution as a threat: http://
datazone.birdlife.org/light-pollution-has-a-negative-impact-on-
many-seabirds-including-several-globally-threatened-species

126–127 Migrating birds are drawn into Twin Towers floodlights:
van Doren, B.M., *et al*, "High-intensity urban light installation
dramatically alters nocturnal bird migration", *Proceedings of the
National Academy of Sciences of the United States of America*,
October 2017

127 Study into common redshanks at Grangemouth: Dwyer, R.G.,
"Shedding light on light: benefits of anthropogenic illumination
to a nocturnally foraging shorebird", *Journal of Animal Ecology*,
November 2012

128 A third of flying insects attracted to street lights will die:
Bruce-White, C., and Shardlow, M., "A Review of the Impact
of Artificial Light on Invertebrates", Buglife – The Invertebrate
Conservation Trust, 2011

128 Global review of insect extinction: Sánchez-Bayo, F., "Worldwide
decline of the entomofauna: A review of its drivers", *Biological
Conservation*, April 2019

128–129 Study of lesser horseshoe bats: Stone, E., *et al*, "Street
lighting disturbs commuting bats", *Current Biology*, July 2009

129 Moths perform a "last ditch evasive manoeuvre": Wakefield, A.,
et al, "Light-emitting diode street lights reduce last-ditch evasive
manoeuvres by moths to bat echolocation calls", *Royal Society
Open Science*, August 2015

129 White lighting attracts more insects: Wakefield, A., *et al*,
"Quantifying the attractiveness of broad-spectrum street lights
to aerial nocturnal insects", *Journal of Applied Ecology*, Vol. 55,
2018, pp.714–722

129–130 Bats display most natural behaviour under red light:
Spoelstra, K., *et al*, "Response of bats to light with different
spectra: light-shy and agile bat presence is affected by white and
green, but not red light", *Proceedings of the Royal Society, B*, May
2017

Notes and References

Chapter 6

page

143–144 Save Tungsten Campaign described the importance of incandescent lighting: statement originally taken from the website of the Association of Lighting Designers' Save Tungsten Campaign in 2013. This has now been superseded by a new campaign, Save Stage Lighting.

149 "Arsenic used to create LEDs", Michael Hulls: see, for example, "LED products billed as eco-friendly contain toxic metals, study finds", *Science Daily*, 11 February 2011

150–151 Howard Brandston on the US incandescent ban: http://concerninglight.com/commentary/

152 "Illumination is atmosphere", municipal council member from Rome: quoted in Povoledo, Elisabetter, "Streetlight Fight in Rome: Golden Glow vs Harsh LED", *New York Times*, 27 March 2017

153 "Cities across Canada are swapping old street lights": "It's early morning all night long: Halifax residents revolt over LED street lights", *The Current*, CBC Radio, 9 May 2017

155 American Medical Association official policy statement: "Human and Environmental Effects of Light Emitting Diode (LED) Community Lighting", American Medical Association, 2016

156 City of Montreal initially planned to instal 4000K street lights: "Montreal drops disputed LED plan after angry protests", *Lux Review*, 20 January 2017

156–157 New York metro's original lighting: Zak, Dan, "The new lighting in Metro is like sitting in a Xerox machine. And it's driving us crazy", *The Washington Post*, 5 April 2017

157 Blue-rich lighting in the Channel Tunnel trains: Brûlé, Tyler, "The Fast Lane: blinded by the light", *Financial Times*, 7 July 2017

157–158 Artificial lighting a "public health hazard": Zielinska-Dabkowska, K., "Make Artificial Lighting Healthier Nature International", *Nature*, 16 January 2018

160 "The history of art is a history of looking at light", James Turrell: quoted in King, Elaine, "Into the Light – A Conversation with James Turrell", *Sculpture*, November 2002

Notes and References

Chapter 7

page

164 Andris Piebalgs on the phasing out of incandescents: "Member States approve the phasing out of incandescent bulbs by 2012", European Commission Press Release Database, 8 December 2008

165 Studies from UK and Sweden on behaviour and energy consumption: "Household Electricity Survey – A study of domestic electrical product use", Intertek Report R66141, May 2012; and Bladh, M., Krantz, H., "Towards a Bright Future? Household use of electric light: a microlevel study", *Energy Policy*, September 2008

167 Sigmar Gabriel proposed using the Ecodesign Directive as a basis to ban incandescents; and "Diving into a sausage machine": Bittner, Jochen, "Ein Schlag auf die Birne", *Die Ziet*, 1 September 2009

167–168 MEPs voted against parliamentary debate over incandescent ban: *ibid*

168–169 "Policies for Energy-Efficient Lighting": *Light's Labour's Lost*, OECD/IEA, 2006

169–170 Report into which Ecodesign requirements could be set for lightingproducts: VITO Final Report, Lot 19: "Domestic lighting". Study for EC DGTREN, October 2009

170–171 Technical briefing for MEPs: https://www.scribd.com/ document/102396085/Phasing-out-Incandescent-bulbs-in-the-EU-Technical-Briefing

171–172 Caroline Lucas on the incandescent ban: "MEP demands ban on old-fashioned light bulbs", press release quoted in *Sutton and Croydon Guardian*, 13 March 2007

172–173 EU legislation implementing measures: Ecodesign Directive, Directive 2009/125/EC

174 "Dimwits!": Booker, Christopher, "Dimwits! Those bright sparks over in Brussels have decided to stop you buying old-fashioned light bulbs", *Daily Mail*, 7 January 2009

174 Comparison between incandescents and CFLs on energy efficiency: Alexander, Ruth, "Why Eco-light bulbs aren't what they seem", *More or Less*, BBC Radio 4, 11 December 2009

Notes and References

174–175 Spectrum Alliance campaign: referred to by Labour MP Sheila Gilmore, UK Parliament House of Commons Adjournment Debate "Incandescent Light Bulbs", *Hansard*, 22 May 2012

176–177 German "action art": "German 'heatball' wheeze outwits EU light bulb ban", Reuters, 18 October 2010; and *Bulb Fiction*, a film by Christopher Mayr (produced by Neue Sentimental Film, Daniel Zuta FilmProduktion, Brandstorm Entertainment and FunDeMental Studios), Austria, September 2011

177 German artists' statement: *Neue Westfälische*, May 2010, http://savethebulb.org/wordpress/artists-against-the-incandescent-lamp-ban

177–178 Documentary exploring the light bulb labyrinth: Mayr, Christopher, *Bulb Fiction* (produced by Neue Sentimental Film, Daniel Zuta FilmProduktion, Brandstorm Entertainment and FunDeMental Studios), Austria, September 2011

184 Legislators closed a loophole: Commission Regulation (EU) 2015/1428, 25 August 2015

185 "Still premature to draw conclusions": European Parliament, Parliamentary Question E-004763/2012, "Proven effectiveness of Commission Regulation (EC) No 245/2009 implementing EcoDesign requirements for non-directional household lamps", 10 May 2012

187 "UN wants to help countries banish incandescent bulbs once and for all": Mother Nature Network, 27 May 2018, https://www.mnn.com/green-tech/research-innovations/blogs/un-light-bulbs-mode-guidelines

188 US$25 million UNEP project: "Transforming the lighting market for sustainable development in Viet Nam", Global Environment Facility, 14 December 2017

188 "400,000 CFLs for Togo": Global Environment Facility, 28 May 2009

188 "Fifty-five countries have joined the en.lighten initiative": UNEP press release, 31 July 2014

188 En.lighten was a multi-million dollar UNEP programme: "The Rapid Transition to Energy Efficient Lighting: An Integrated Policy Approach", UNEP policy brochure, 2012; and https://www.

unenvironment.org/news-and-stories/story/brighten-making-
switch-efficient-lighting

188–189 En.lighten became part of United4efficiency: "Simple Ways
to Achieve Energy Efficiency", U4E brochure, 2017

189–190 Recall of lamps involved in fires: https://www.cpsc.gov/
content/trisonic-compact-fluorescent-light-bulbs-recalled-due-
fire-hazard

190 "Texas tells Feds": *Illinois Review*, 20 June 2011

190–191 Howard Brandston: testimony to the Senate Energy
Committee, Washington DC, 10 March 2011

193 "Intellectually dishonest", Dr John Lincoln: personal
communication to author, August 2018

196 Minna Gillberg: quoted by Balksjö, Jessica, in "Kvicksilvret i
lampor går rätt ut nature", *Svenska Dagbladet*, 20 November
2011

Chapter 8

page

198–199 Internet discussions about the incandescent ban: comments
below https://www.youtube.com/watch?v=xO5lGpFGcJY

199 "Save money": comments below http://blog.brian-fitzgerald.net/
blog/2006/07/06/ban-the-bulb/

200 Book about an incandescent light bulb on death row: Leahy,
Michael Patrick, *I, Light Bulb*, Broadside Books, 2011

200 Telegraph headline: Gray, Richard, "Official responsible for light
bulb ban is a former communist", *Telegraph*, 12 September 2009

200 "What the light bulb ban has done is to treat consumers like
idiots": Clark, Ross, "Lightbulb ban was about money not the
environment", *Daily Express*, 28 January 2014

200 *Daily Mail* gave out 25,000 incandescent light bulbs: Derbyshire,
David, "Lights go out as Britain bids farewell to the traditional
bulb despite health fears about eco-bulbs", *Daily Mail*, 6 January
2009

201 *Guardian* on the incandescent ban: Randerson, James, "Good
riddance to incandescent lightbulbs", *Guardian*, 7 January 2009

201   Change to CFLs is a "no-brainer": Tran, Mark, "Incandescent rage as lights go out on old 100w bulbs", *Guardian*, 31 August 2009

201   "Wasteful and unnecessary technology": Monbiot, George, "Drastic action on climate change is needed now – and here's the plan", *Guardian*, 31 October 2006

201   *Daily Mail* warning: Belgado, Martin, "An energy saving bulb has gone – evacuate the room now!", *Mail Online*, 6 January 2008

201   *Telegraph* flags up health concerns: Gray, Louise, "Last chance to buy old-fashioned bulbs", *Telegraph*, 1 September 2012

202   *Guardian* dismissed health concerns: Randerson, James, "Good riddance to incandescent lightbulbs", *Guardian*, 7 January 2009

203   "Our aim is that every householder installs low energy light bulbs", Gordon Brown: Prime Minister's speech on the environment to the Foreign Press Association, 19 November 2007

203   "Ambitious efficiency criteria for lights need to be introduced", Sigmar Gabriel: quoted in "Call to ban inefficient lightbulbs in EU", *EU Observer*, 26 February 2007

203   "I will immediately sign a law that begins to phase out incandescent light bulbs", Barack Obama: "Real leadership for a clean energy future", speech, 8 October 2007

203   "This is yet another example of ill-thought-out legislation made by the EU before they have got all the facts", Paul Nuttall MEP: quoted in "Low-energy bulb warning", *Rochdale Online*, 15 May 2014

203   "Rather than forcing people to switch by removing their choice we should be using the right incentives to encourage change", Martin Callanan MEP: quoted in "Official responsible for light bulb ban is a former communist", *Telegraph*, 12 September 2009

203   "The American people want less government intrusion into their lives, not more, and that includes staying out of their personal light bulb choices", Michele Bachmann: statement in support for Light Bulb Freedom of Choice Act, 1 March 2011

203   "Remember, new 'environment friendly' light bulbs can cause cancer. Be careful – the idiots who came up with this stuff don't

Notes and References

care", Donald Trump: tweet by @realDonaldTrump, Twitter,
17 October 2012

204 Labour MP Sheila Gilmore in Westminster: UK Parliament
House of Commons Adjournment Debate "Incandescent Light
Bulbs", *Hansard*, 22 May 2012

204 Contesting the Single Lighting Regulation: Regula Rytz and
Yvette Estermann quoted in "Grüne und SVP fordern: Bundesrat
soll für 'gesundes licht' sorgen", *Nau.ch*, 9 June 2018

Chapter 9

page

209–210 Professor Harald Haas, TED talk, July 2011: https://www.ted.
com/talks/harald_haas_wireless_data_from_every_light_bulb

210 Professor Harald Haas, TED talk, September 2015: https://
www.ted.com/talks/harald_haas_a_breakthrough_new_kind_of_
wireless_internet

211 New French law for the protection of darkness: "Arrêté du
27 décembre 2018 relatif à la prévention, à la réduction et à la
limitation des nuisances lumineuses", NOR: TREP1831126A,
Article 3, Section I

216 Industry definition of human centric lighting: https://www.
lightingeurope.org/human-centric-lighting

217 "We can't develop human centric lighting until we know what
impact light has upon human biology across the day and night
cycle", Professor Russell Foster: quoted in Molony, Rau, "Too
early to start human-centric lighting", *Lux Review*, 26 March
2018

217–218 Warning of misunderstanding of LEDs: Zielinska-
Dabkowska, Karolina, "Human Centric Lighting – The New X
Factor?", *arc Magazine*, 12 December 2018

219 "Bat-friendly lighting": Signify press release, 5 June 2018

220–221 Günther Oettinger acknowledged the potential problems
for light-sensitive people: quoted in press release from Struan
Stevenson MEP, 11 June 2013

221 Nobel Prize suggestion: Zielinska-Dabkowska, Karolina,

"Human Centric Lighting – The New X Factor?", *arc Magazine*, 12 December 2018

221 "Such stubborn fondness": Brox, Jane, *Brilliant: The Evolution of Artificial Light*, Souvenir Press, 2010

*

This has been a journey into unfamiliar territory and some very strange terrain, and I'm grateful to those individuals from around the world who have beaten a path before me, gathered information and shared it online.

I am not necessarily endorsing the information, views or politics expressed on the following websites, but I share the authors' conviction that there is a story that needs to be told.

www.gluehbirne.ist.org (in German) by Peter Stenzel, Austria
https://lightmareaustralia.weebly.com/ by Borek Puza, Australia
http://users.skynet.be/fc298377/EN_argument.htm by Rik Gheysens, Belgium
freedomlightbulb.blogspot.com by Peter Thornes, Ireland
https://greenwashinglamps.wordpress.com/ by Inger G. Nordangård, Sweden

Acknowledgements

Thank you to everyone who contributed – whether mentioned in the book or behind the scenes – for information and explanation, for engaging with my relentless questions, for checking drafts and for trusting me to tell your stories.

Special thanks to Dr John Lincoln and all involved in the charity LightAware (www.lightaware.org).

I'm grateful for the work of the late Catherine Hessett, who contributed so much through her involvement with Spectrum Alliance. I'm sad that I never met her, but I have drawn on her letters and writings throughout my research for this book, and her legacy inspires me still.

Thank you to all at Saraband, especially to Sara Hunt for "getting it" when so many around me didn't, and to Craig Hillsley for shaping it with such clarity.

For space to write through a year of upheaval, my thanks to Sara M for precious days in Portobello, Leo for a much needed sofa-with-dogs, and to Joon for the sanctuary of an orchard studio and my first green woodpeckers.

Acknowledgements

On a personal level, I'm grateful to all who helped me through the darkest parts of those first light years. My thanks to all who could hear me, especially Carolyn and Per for changing the light bulbs and putting the kettle on.

I felt upheld by Polmont Quaker Meeting throughout this work. Special thanks to Jessica for listening, Cath for guidance and Maureen for pointing out that this is a human rights issue.

I'm grateful to Mum, Dad, Steve and all my family, for constant support and encouragement.

And thanks to Rob, for everything.

Index

Index

Index

Index

Edinburgh, 8, 27, 55, 57, 58, 144: Edinburgh University, 209; George Street, 27, 57; Heriot-Watt University, 54; Princes Street Gardens, 197; Scottish National Portrait Gallery, 141; Waverley Station, 32

Edison, Thomas, 1, 66, 189

Edwardian era, 26

Einstein, Albert, 57

electricity, 13, 16, 24, 26, 58, 65–66, 164–165, 169, 171, 178, 182, 226–227, 232

electroluminescence, 77

electromagnetic field (EMF), 75–77: *see also* electromagnetic radiation/spectrum

electromagnetic radiation/spectrum, 53, 57–64, 78, 80, 97, 100–101, 110, 114, 210, 232: microwaves, 57, 63; radio waves, 60, 63, 210; spectral distribution, 68, 78, 191, 218; visible spectrum, 63, 68–71, 84, 95, 100, 121, 151, 218; *see also* blue light; electromagnetic field; frequency; gamma rays; infrared; photons; ultraviolet radiation; waves/wavelength; X-rays

employment tribunal, 47–48

England *see* UK *and individual places*

en.lighten, 187–188: *see also* Global Environment Facility; United4efficiency; United Nations Environment Programme

Environmental Health Perspectives, 92–93

environmentalists/ism, 13, 23, 171, 180–181, 198–200, 205, 227–229: "green" politics, 20, 50

epilepsy, 98, 111

Essex University, 97

Estermann, Yvette, 204

Europe, 148, 164, 174, 179–180, 183, 185–186, 189, 192, 194, 200, 203–204: Brussels, 1, 50, 83, 164, 179; Conservative politicians (in Europe), 168, 204; European Commission, 1, 164–166, 171, 174, 200, 220–221; European Parliament, 13, 49–50, 166, 167–168, 170–171, 183–184, 185, 203–204; European Union (EU), 1, 12–13, 22, 37, 49–50, 88, 111, 143, 164–169, 171, 172, 174–175, 177, 180–181, 182–186, 192–193, 196, 200, 203–205, 213, 220–221, 228; Labour politicians (in Europe), 204; Scientific Committee on Emerging and Newly Identified Health Risks (SCENHIR), 111–112, 183–184; Scientific Committee on Health, Environmental and Emerging Risks (SCHEER), 112–113, 193; *see also individual countries*

evolution, 102–103, 119

eyes, 41, 68–69, 89–96, 102–104, 113, 191–192, 205, 226: colour blindness, 60; cones, 69, 103; *International Review of Ophthalmic Optics*, 94–95; intrinsically photosensitive retinal ganglion cells (ipRGCs), 103, 105, 106, 109, 114, 217; macular degeneration, 20, 94, (age-related macular degeneration, AMD), 113; ophthalmologist/optician, 41, 81, 94, 232; *Optician*, 94; photochemical retinopathy, 112; retinal disease, 111; rods, 69, 82, 83, 103

Exeter University, 130: The Environment and Sustainability Institute, 133

fairy lights, 138

Falkirk, 138, 156

Faraday, Michael, 58, 232

fibromyalgia, 111

Fife, 8

Finland *see* Nordic

fire(light), 26, 28, 65, 66, 78, 125, 138, 147, 150, 183, 190, 208, 234

fire risk, 189–190

fireworks, 125

fish: cookie–cutter shark, 125; dragonfish, 125–126; phytoplankton, 125, 137; octopus, 126; squid, 125

fishing, 123–124: jigging, 125

flicker, 3, 12, 40, 43–44, 53, 97–99, 111–114, 142, 205, 222, 223: *see also* electromagnetic radiation/spectrum; frequency; waves/wavelength

fluorescent light, 3, 8, 15, 25, 27, 31, 35, 42, 44, 72, 76–77, 83–84, 89, 96, 110–111, 115, 138–139, 151–152, 161, 169, 175,

Index

192, 210, 213, 223: CFLs (compact fluorescent lights), 4, 7, 10–13, 15–16, 19–21, 23, 28–30, 36, 49, 56–57, 59, 70, 81–87, 90, 94–96, 98, 99–101, 106, 108–109, 110–112, 138, 149–151, 157, 161, 165, 169–174, 176–179, 180, 181–183, 186–192, 200–202, 206, 213–215, 225, 229; double envelope CFL, 85–86, 165; single envelope, 85

food chain, 16, 125, 131, 136, 188, 214

Forth, 31, 127: Firth of, 8

Foster, Professor Russell, 217

Fox, George, 8

France, 89, 91–92, 211

frequency, 30, 43, 62–63, 82, 97, 98, 109, 110, 111, 121, 122, 130, 205, 209, 210, 218: *see also* electromagnetic radiation/ spectrum; flicker; waves/wavelength

Friends of the Earth, 15, 202: *see also* environmentalists/ism

Gabriel, Sigmar, 167, 203

gamma rays, 64: *see also* electromagnetic radiation/spectrum; frequency; waves/ wavelength

Gaston, Professor Kevin, 130–132

Gateshead, 1

General Electric, 66–67

Germany, 167, 168, 176–177, 178: Berlin, 13, 178; East Germany, 168; German Radiation Protection Agency, 101

Gillberg, Minna, 196

Gilmore, Sheila, 204

Girl in the Dark see Lyndsey, Anna

glasses, 42: blue-blockers, 40; sunglasses, 157

glare, 4, 14, 42, 90, 92, 97, 121, 153–154

Glen Affric, 117–118: Loch Affric, 118

Global Environment Facility (GEF), 187–189: *see also* en.lighten; United4efficiency; United Nations Environment Programme

Grangemouth oil refinery, 8, 127, 223

Green Party *see* UK government and politics

Greenpeace, 13, 15, 202, 178: *see also* environmentalists/ism

"greenwashing", 200, 228–229

"Goldilocks light", 233–234

Google, 12, 15, 16, 20, 22, 37, 78

Haas, Harald, 209–210

halogen gas, 67

halogen light, 9, 17, 20, 32, 56, 57, 62, 66–67, 112, 192, 204, 219, 204, 219

Hansard, 166

Harvard Medical School, 106

Hattar, Samer, 109–110

Hawk, Professor John, 82–83, 176

heart, 80, 87, 97: heart disease, 106

Heatballs *see* Rotthaeuser, Siegfried

Hertz, Heinrich, 60

Hessett, Catherine, 175: *see also* Spectrum Alliance

hoarding, 175–176: *hamsterkäufe*, 176

hormones, 22, 46, 103–104, 106, 107, 108, 137, 232: *see also* melatonin

Hulls, Michael, 147–150, 176: *Castor and Pollox*, 148; *Lightspace*, 148

human centric lighting, 216–219

human rights, 50–52, 54–55, 197–198, 202

hygge, 186

ignorance, 27, 80–82, 228

I, Light Bulb see Leahy, Michael Patrick

incandescent lighting, 1, 9, 12–13, 15, 22–23, 26, 27, 44, 49–50, 56–57, 64, 66–67, 70, 72, 81, 83, 84–85, 86, 88–89, 94–96, 110, 111, 112, 139, 141, 143–144, 146, 147, 148, 149, 150, 162–196, 198–206, 209, 210, 218, 220–221, 226–227, 234: ban on bulbs, 162–196; rough service lamps, 184; specialist lamps, 184; squirrel cage lamps, 184

India, 13, 16, 177

infrared, 63, 110, 218: *see also* colours; electromagnetic radiation/spectrum; frequency; waves/wavelength

insects, 127–129, 232: butterfly, 130; caterpillar, 130; cockroach, 128; dog whelk, 133–134, 137; dung beetle, 120; earthworm, 128; earwig, 128; firefly, 128, 137; midge, 128; moth, 128, 130; spider, 25; woodlouse, 128

Index

Index

Index

Index

Index